Electronics Con___
Labs, and Projects:

For Media Enthusiasts, Students, and Professionals

Electronics Concepts, Labs, and Projects:

For Media Enthusiasts, Students, and Professionals

Alden Hackmann

Hal Leonard Books
An imprint of Hal Leonard Corporation

Published in 2014 by Hal Leonard Books
An Imprint of Hal Leonard Corporation
7777 West Bluemound Road
Milwaukee, WI 53213

Trade Book Division Editorial Offices
33 Plymouth St., Montclair, NJ 07042

Printed in the United States of America

Book design by Bill Gibson

Library of Congress Cataloging-in-Publication Data

Hackmann, Alden.
 Electronics concepts, labs, and projects : for media enthusiasts, students, and professionals /
Alden Hackmann.
    pages cm
Includes bibliographical references and index.
 ISBN 978-1-4803-4243-9
 1. Electronics--Textbooks. 2. Electronics--Laboratory manuals. I. Title.
 TK7816.H335 2014
 621.381--dc23
                       2014025455

www.halleonardbooks.com

To Cali, who has made everything possible

Contents

Preface: Why We Are Here

Without electronics, the audio world would be essentially nowhere. For recording, we would be using wax cylinders. For live sound, we would be using a megaphone. The roots of the history of electronics stretch back at least 2,600 years to ancient Greece. Many scholars and inventors over the years have added to the knowledge: Benjamin Franklin, J. J. Thomson, Nikola Tesla, Thomas Edison, Alessandro Volta, Georg Simon Ohm, and Gustav Kirchhoff, to name a few. (The fact that there are no women on this list doesn't suggest that women didn't have the ability to contribute to the field. It's a regrettable part of the history of science that few women were in the right place at the right time to apply their insights to the problems. That's a subject for an entirely different book.) It could be argued that these people were brilliant scientists, and they certainly were, but they still had to put their pants on one leg at a time like everyone else. More importantly, they built on what had gone before: while their contributions were revolutionary, it took all of them and many more to bring us to the modern electronic world of the Neumann U87 microphone, the Universal Audio 6176 preamp-compressor, the revered SSL console, the iPhone, and the soundboard tapes recorded by Betty Cantor-Jackson.

Whether you are learning electronics in a class or studying it on your own, you may be wondering what the goals of your studies might be. You will encounter many situations as an audio professional where knowledge of electronics will be helpful. Even if you never open up a piece of gear, an understanding of the concepts of the electronics will add to your ability as an audio professional. This text will teach you the basics of troubleshooting gear: electrical outlets with low voltage, bad cables, faulty microphones, malfunctioning preamps, consoles with bad power supplies, flaky power amps, speakers with intermittent output, guitar amps that shock you, and so on. You can expect to encounter all of these things and more during your career. Each chapter begins with a list of the objectives to be learned. In the education field, we call these competencies, meaning that "when students are done with this part of the course, we can expect that they will be competent with this set of skills."

Most electronics books are written by electrical engineers, who like to look at the subject from a math approach. I've done my best to just give you the math you need, without getting it too complicated. To get the most out of this book, it's good to use a little bit of math; it helps cement the information into place and make it tangible. I'm well aware that on average, audio people don't usually list math as their favorite subject. If you can use a basic calculator to add, subtract, multiply, and divide, then you can do the math in this book. Occasionally you need the log key, but only for the decibel equation. One of the appendices shows my method of solving the math questions that will let you break them down into logical steps. If you really, really don't like math, you can skip the calculations. Just be aware that you'll then need to take my word for a few things.

In order to figure out how to diagnose these difficulties and be safe while doing so, you need to know the basics of electricity and electronics. I don't expect you to design a new tube preamp on the back of your paper napkin during lunch, but learning and putting into practice the information in this text will let you know where to set your multimeter to figure out whether the electrical outlet is working, or if the fuse is blown, or whether there is phantom power being supplied to the microphone. With these skills, you can show yourself to be a valuable member of a production team, and you can start to fix your own gear.

There will be frustrating moments along the way, and you probably will make some mistakes that make the problem worse rather than making it better. This is called experience, and there isn't an electronics tech in the world who hasn't done this. Despite these moments of annoyance, or in fact because of them, the skills you learn here will lead you down the path to a better understanding of how all of this gear functions, how to care for it so that it will last longer and behave better, and what tools to buy if you want to start working more seriously on your own gear or someone else's.

One challenge in learning electronics is that there is a body of knowledge that is all interconnected. This makes it hard to know where to start. You may find that there are terms and concepts that come up in your reading that you don't understand yet. Don't let this discourage you—rest assured that everything will be explained in time. It's not reasonable to expect all of this to sink in the first time. Keep at it, keep reading, keep experimenting, and keep learning.

Good luck, and a safe journey!

Alden Hackmann

Acknowledgements

Many people have helped me with this book. In particular, my thanks go to:

- All of my students who have taught me so much.
- Steve Barsotti.
- Bill Gibson.
- Jim Rivers, alway generous with his knowledge.
- David Christensen.
- The "band beyond description" who kept me in the zone through the writing process.
- Justin and Megan, who always believe and inspire.
- Kiran, for introducing me to the gateway to coffee.
- Debbie and Amanda at Commuter Comforts, all my baristas at the downtown Starbucks, Ricky at Le Panier, Kathy at the UW, Jessica at the Coffee Dock, and Amy and all the friendly people at the Cup and Muffin: this book would not have happened without you.
- Casey Burns, for introducing me to the joys of the GoPro camera.
- The extended Hackmann and Reil families, for your great support.
- Katherine, especial thanks for manuscript reading and critique.

And as in the dedication, thanks and love always to my darling Cali.

Author's Note: Legal Liability

Working with electricity in any form can be dangerous in a number of ways. I will do my best to advise you of the dangers and tell you how to avoid the hazards. I cannot emphasize strongly enough that electricity is potentially lethal, and you must treat it with extreme respect in order to stay alive and uninjured. Neither I nor the publisher, Hal Leonard, can accept any responsibility or liability for anything you do that may result in property damage, injury, or death.

Introduction: What You Need

To understand the basics of electronics, you don't really need anything except this book. There's no reason that you can't just sit in a comfortable armchair, read the book cover to cover, and learn everything I've written about. With that said, I've been teaching this material to audio students for over 10 years, and the vast majority of them say that it's the hands-on lab work that brings to life the words in this book and in my class presentations. I strongly recommend that you obtain a few basic supplies for the journey ahead. If you don't practice as you go along, you won't really have anything except theoretical knowledge when you are done reading.

If you are not already familiar with the prefixes **M, k, m, μ, n**, and **p**, I suggest that you consult Appendix A. These are the standard prefixes for electronics, and you will be using them constantly. A prefixes table is included in the supplemental materials on the DVD.

I also recommend using a circuit simulator, either as a program on a computer or as an app on a smartphone. These are discussed in Appendix D, and one is included on the DVD. The simulator lets you build circuits on your screen and observe how they act.

All of the media examples can be found on the accompanying DVD-ROM and online at www.halleonardbooks.com/ebookmedia/HLitem#. E-book readers can download all the content that is included on the DVD-ROM that accompanies the print edition.

Supplies Lists

Below are lists of pieces of equipment and supplies that are mentioned in this book. If you want to get right down to reading about electronics, by all means you should skip ahead to the first chapter. You can come back later when you have a better idea of what you are interested in working on first. There are three lists: the Soldering Supplies List for building the projects, the Exercises Supplies List for doing the hands-on exercises at the ends of the chapters, and the Project Supplies List of the materials for the projects. Each set of exercises and project instructions also includes a list of supplies specific to that exercise or project. Every item is described in detail after the lists, with an explanation about what to look for and where to get it.

You don't need to go out and get all these materials right away. I have assembled these lists so that you don't have to go through chapter by chapter and project by project to figure out what you need. There is a roster of sources at the end, with details of the vendors mentioned throughout the materials lists.

The Soldering Supplies List

Safety glasses

Soldering station or soldering iron

Solder

Insulated solid core wire (20, 22, or 24 gauge)

1/4" jack or similar jack with solder lugs on it for soldering practice

Desoldering pump

Desoldering braid

Helping Hands jig

Panavise or similar vise

Reamer/scraper tool

Wire strippers

Flush cutting wire cutters

Needle nose pliers

Side cutting wire cutters

Craft knife (X-acto or similar)

Coddington magnifier

The Exercises Supplies List

Digital multimeter (DMM)

Polyswitch 250mA self-resetting breaker

Calculator

Solderless breadboard

9V battery

9V battery clip

100Ω resistors (100 ohms)

470Ω resistors (470 ohms)

560Ω resistors (560 ohms)

$1.0k\Omega$ resistors (1000 ohms)

$4.7k\Omega$ resistors (4700 ohms)

$10k\Omega$ resistors (10,000 ohms)

$22k\Omega$ resistors (22,000 ohms)

$47k\Omega$ resistors (47,000 ohms)

$75k\Omega$ resistors (75,000 ohms)

$100k\Omega$ resistors (100,000 ohms)

$220k\Omega$ resistors (220,000 ohms)

$470k\Omega$ resistors (470,000 ohms)

$10k\Omega$ linear taper potentiometer (10,000 ohms)

$10k\Omega$ audio taper potentiometer (10,000 ohms)

1N4002 diodes

Green LEDs

Red LEDs

2N2222 NPN transistors

0.33µF polyester capacitor (0.33 microfarad)

1µF polyester capacitor (1 microfarad)

10µF electrolytic capacitor (10 microfarads)

100µF electrolytic capacitor (100 microfarads)

1000µF electrolytic capacitor (1000 microfarads)

Stopwatch, watch with a second hand, or equivalent

GFCI socket or adaptor

Jumper wires with alligator clips

Transformer, 120V primary, 9-18V AC secondary

Optional: grab bag of resistors

Optional: grab bag of capacitors

Optional: grab bag of potentiometers

Optional: grab bag of LEDs

Optional: dead piece of audio gear

The Project Supplies List

Three TS connectors

Two TRS connectors

One female XLR connector

One male XLR connector

Unbalanced heavy-duty cable

Unbalanced light-duty cable

Star quad cable

Balanced heavy-duty cable

Piezo disc

Specifications and Sources for Materials

The details and specifications for the soldering, exercises, and project materials are given here, followed by a listing of places to get them. These specifications are intended to guide you so that you have a better idea of what to look for in the dizzying array of materials that are available to you. My intent in being so specific is not to restrict your choices, but to make it easier to choose and to follow along with the exercises and projects.

For most of the materials, I've given item numbers from several companies for your convenience. I can't guarantee that any particular item will remain available. DigiKey and Techni-Tool have been helpful to me with suggesting alternatives when a particular item has become obsolete. It is worth noting that if you are ordering the materials for the Exercises Supplies List from DigiKey, there are several items on the other lists that are also available from DigiKey. If you are placing a DigiKey order, it's easiest just to get everything at once.

The Soldering Supplies List Details

- **Safety glasses:** These are required if you want to protect your eyesight. You can get any style that will protect your eyes from the sides as well as the front. [Item 278IE001 from Techni-Tool; any hardware store; bifocal models are available also from The Woodworker's Store and others.]

- **Soldering station:** As mentioned in the chapter on soldering, get a good one with a temperature-controlled tip, preferably ESD-safe. The Weller WES51 stations are a good choice, though expensive as a starting point. [Item WES51-ND from DigiKey.]

- **Solder:** A pound of wire solder will last you a very long time. Either 60/40 or 63/37 (40% lead or 37% lead) is fine, with rosin activated (RA) or rosin mildly activated (RMA) flux, about 0.025" (22 or 21 gauge). [Item KE1106-ND from DigiKey.]

- **Insulated solid core wire:** 20, 22, or 24 gauge. [Item C2117B-100-ND from DigiKey.]

- **1/4-inch jack** or similar jack with solder lugs on it: This is for soldering practice. Anything with solder lugs will work, but a simple panel mount mono or stereo jack is preferred. These are widely available at guitar stores that sell parts for working on guitar pedals. [Item SC1094-ND from DigiKey or similar.]

- **Desoldering pump** ("solder sucker"): This is a tool for sucking up molten solder. Paladin makes a nice one that's reasonably priced (Model 1700), but they pretty much all work the same. [Item PAL1700-ND from DigiKey.]

- **Desoldering braid:** This is the other way of removing solder. [Item EB1091-ND from DigiKey.]

- **Extra hands positioning tool:** This is a device that looks like a pair of alligator clips that can be positioned wherever you want to hold two pieces to be soldered together. Some come with magnifying glasses, which I always take off because they get in the way but you may find useful. [Item 890PR230 from Techni-Tool; Radio Shack.]

- **Panavise** or similar: This easily positioned vise is invaluable for holding connectors, wires, and circuit boards. This vise is a bit expensive, but well worth it for its capabilities. [Item 301PV-ND from DigiKey.]

- **Reamer/scraper tool:** The reamer is used for unbraiding shields on audio cable. It's bent and pointed on one end with a sharp scraping blade on the other. [Item 388SO022 from Techni-Tool; Radio Shack.]

- **Wire strippers:** Get the strippers with the sets of matched holes that fit at least wire sizes 14–20. Don't get the cheap adjustable ones with the sliding nut. Techni-Tool's house-brand strippers are decent. Klein is reasonable and widely available at hardware stores. Wiha are the best, but they are expensive. If you are doing work with really small wires, you'll probably end up wanting another stripper for the smaller sizes. I have yet to find a single stripper that covers the range of wires I work with. I'm always either at the small end of one set or the large end of the next. [Item 758PL0031 from Techni-Tool.]

- **Flush cutting pliers:** These are wire cutters specifically for cutting off the excess wire from circuit boards. You can use standard side cutters, but the flush cutting pliers give a much cleaner result. My favorites are Excelite 170M. If you get flush cutters, don't use them as regular wire cutters. They will last for years if you don't abuse them, but they aren't designed for cutting through thick cables. They are also great for trimming the packing material in cable, but only if they are kept sharp. [Item 170M-ND from DigiKey.]

- **Needle nose pliers:** Budget needle nose pliers from a hardware store are fine. Look for a pair with a fairly fine tip, maybe a little wider than a toothpick, as what a hardware store calls needle nose pliers can be too large. The ones from electronics supply houses are comparatively quite expensive. Jewelry supply stores and craft stores have great selections. Wiha brand are really nice, but expensive. [No item number listed because electronics supplies are outrageously priced, and I prefer to pick them up and look at them.]

- **Side cutting wire cutters:** These are also called "dikes," short for "diagonal cutters." They are good for cutting heavy wire and cable. Wire strippers also have wire cutters, but dikes are heavier. Hardware-store wire cutters are a good choice here. Wiha's are really nice and will last a lifetime, but they are expensive. [Item 431-1052-ND from DigiKey.]

- **Craft knife** (X-acto or similar): Used for stripping insulation from the outside of audio cables. These are widely available at a hobby shop, art supply store, or hardware store. Alternatively you can use a utility knife with replaceable blades for this job. [Item X-3201 from X-acto or similar.]

- **Coddington magnifier:** This is a small 10X illuminated magnifier. It's not required, but I have found it invaluable for inspecting solder joints and reading the tiny print on capacitors and IC chips. [Item 06546840 from MSC Industrial.]

The Exercise Supplies List Details

- **Digital multimeter (DMM):** The meter needs to be able to read AC voltage, DC voltage, DC current, resistance, and continuity. An audible beep or buzzer on the continuity tester setting is highly recommended, and most meters these days have this feature. Pretty much any DMM on the market will do the job, but make sure that it reads DC current; not every meter does. A higher-quality meter such as a Fluke will last longer and possibly have more features, but it will be much more expensive. A $20 meter will work for you, at least for a while. Recent years have seen a decline in the quality of entry-level multimeters, so if you pay a minimal amount, you can expect to be replacing it pretty soon. In particular, the meters from a well-known chain of electronics supplies should be avoided if possible. There are kits available, but I don't recommend them. [Electronics stores, hardware stores, online sources.]

- **Polyswitch 250mA self-resetting breaker:** You can modify your DMM by replacing the fuse with this nifty device. If there is too much current, the breaker opens and stays that way until the danger has passed. The breaker waits a little while to cool down, then resets itself. [Included in the Supplies Kit from Bracken Creek, or item TRF250-120-ND from DigiKey.]

- **Calculator:** You need a calculator with a log key. I'm old school, so I prefer calculators with buttons, but the calculator app on a smartphone works fine.

- **Solderless breadboard:** This is a plastic block with holes containing spring-loaded strips that grip the wire leads of components to make connections. I prefer the one mounted on a board with three binding posts, but an unmounted one is fine, as shown in the figures. The mounted one comes with a set of preformed jumper wires. [Unmounted item 438-1045-ND, mounted item 438-1047-ND from DigiKey; online sources.]

- **9V battery:** Widely available. [Included in the Supplies Kit from Bracken Creek.]

- **9V battery clip:** Widely available. [Included in the Supplies Kit from Bracken Creek, or item BS6I-HD-ND from DigiKey.]

- **Resistors:** All resistors should be 1/4 watt with 5% tolerance. Packages of five resistors are typical. All of these resistors are included in the Supplies Kit from Bracken Creek. If you prefer to order from DigiKey, the item number is shown for each.
    - 100Ω resistor (100 ohms) [item 100QBK-ND from DigiKey].
    - 470Ω resistor (470 ohms) [item 470QBK-ND from DigiKey].
    - 560Ω resistor (560 ohms) [item 560QBK-ND from DigiKey].
    - 1.0kΩ resistor (1000 ohms) [item 1.0KQBK-ND from DigiKey].
    - 4.7kΩ resistor (4700 ohms) [item 4.7KQBK-ND from DigiKey].
    - 10kΩ resistor (10,000 ohms) [item 10KQBK-ND from DigiKey].
    - 22kΩ resistor (22,000 ohms) [item 22KQBK-ND from DigiKey].
    - 47kΩ resistor (47,000 ohms) [item 47KQBK-ND from DigiKey].
    - 75kΩ resistor (75,000 ohms) [item 75KQBK-ND from DigiKey].
    - 100kΩ resistor (100,000 ohms) [item 100KQBK-ND from DigiKey].
    - 220kΩ resistor (220,000 ohms) [item 220KQBK-ND from DigiKey].
    - 470kΩ resistor (470,000 ohms) [item 470KQBK-ND from DigiKey].

- **10kΩ linear taper potentiometer** (10,000 ohms): 1/10 watt or larger, circuit board or panel mount—look for a potentiometer with an easy way to tell where the shaft is positioned, i.e., a flat or a slot in the shaft. [Included in the Supplies Kit from Bracken Creek, or item 987-1308-ND from DigiKey.]

- **10kΩ audio taper potentiometer** (10,000 ohms): 1/10 watt or larger, circuit board or panel mount—as for the linear potentiometer, look for one with a flat or slot in the shaft. [Included in the Supplies Kit from Bracken Creek, or item 987-1307-ND from DigiKey.]

- **1N4002 diodes:** These are the industry standard diode. Packages of 5 or 10 are typical. [Included in the Supplies Kit from Bracken Creek, or item 1N4002FSCT-ND from DigiKey.]

- **Green LEDs:** Get at least 10 of these, because they don't last forever in the stress of the exercises. [Included in the Supplies Kit from Bracken Creek, or item 1080-1127-ND from DigiKey.]

- **Red LEDs:** We want two colors because we will use them to distinguish easily between different parts of the circuit. Again, get at least 10 or 15 of these, because they don't last forever. [Included in the Supplies Kit from Bracken Creek, or item 67-1648-ND from DigiKey.]

- **2N2222 NPN transistors:** This is a typical NPN audio transistor. Get a few in case you blow some up (which is normal when one is first working with transistors). [Included in the Supplies Kit from Bracken Creek, or item P2N2222AGOS-ND from DigiKey.]

- **Capacitors:** Low-voltage (10V or more) capacitors are fine, because you aren't going to be using them with a voltage higher than 9V from a battery. Packages of five capacitors are typical. All of these capacitors are included in the supplies kit from Bracken Creek. If you prefer to order from DigiKey, the item number is shown for each.
    - 0.33μF polyester capacitor (0.33 microfarad) [item P4549-ND from DigiKey].
    - 1μF polyester capacitor (1 microfarad) [item EF2105-ND from DigiKey].
    - 10μF electrolytic capacitor (10 microfarads) [item P5134-ND from DigiKey].
    - 100μF electrolytic capacitor (100 microfarads) [item P5138-ND from DigiKey].
    - 1000μF electrolytic capacitor (1000 microfarads) [item P5127-ND from DigiKey].

- **Stopwatch,** watch with a second hand, or equivalent: Most phones have a second counter or a stopwatch function. A dedicated stopwatch is not needed.

- **GFCI socket or adaptor:** If you have a GFCI wall socket available where you are working, use that. If you don't, you can get an adaptor at the hardware store for $15 or $20. [Also available online.]

- **Jumper wires with alligator clips:** You'll need at least four of these, preferably of different colors so you can tell them apart easily. [Radio Shack, Electronic Goldmine.]

- **Transformer,** 120V primary, 9-18V AC secondary: Any AC-to-AC transformer with an output of 9V to 18V will work. Check that the output is AC, as many of these "wall warts" are designed for electronics that run on DC. [Radio Shack, Electronic Goldmine.]

- **Optional grab bags** of resistors, capacitors, potentiometers, LEDs: I suggest getting some grab bags of components so you can work with sorting through them and identifying them. These grab bags are usually inexpensive and contain a variety of components in a dizzying array of values and formats. [Electronic Goldmine, other online sources.]

- **Optional dead pieces of audio gear:** I also suggest picking up a piece of audio gear from eBay or Craigslist or your local music gear store. You're looking for something nonfunctional, so it's cheap. You'll use them for looking around, measuring voltages and resistances, and identifying components.

The Project Supplies List Details

- **TS connector:** model Neutrik NP2C. Three are needed: two for the TS cable, one for the piezo cable. [Redco, Full Compass.]

- **TRS connector:** model Neutrik NP3C. Two needed for the TRS cable. [Redco, Full Compass.]

- **XLR female connector:** model Neutrik NC3FX. One needed. [Redco, Full Compass.]

- **XLR male connector:** model Neutrik NC3MX. One needed. [Redco, Full Compass.]

- **Unbalanced cable, heavy duty.** One piece of desired length for the TS (guitar) cable. There are a range of cable types available for this application. Here are some good options: Redco TGS-HD instrument cable, Canare GS-6 instrument cable, Mogami W2524 instrument cable. [Redco, Full Compass.]

- **Unbalanced cable, light duty.** One piece of desired length for the piezo transducer, where it's better to keep the cable lightweight and down to a length of 6' or less. Here are some good options: Canare GS-4 mini instrument cable, Mogami W3219 miniature instrument cable. [Redco, Full Compass.]

- **Star quad cable.** One piece of desired length for a microphone cable. There are many cables available for this application. Here are some proven options: Canare L-4E6S quad mic cable, Redco TGS-QD quad mic/line cable, Mogami W2534 quad mic cable. [Redco, Full Compass.]

- **Balanced cable.** One piece of desired length for a balanced signal cable. There are many cables available for this application. Here are some proven options: Canare L-2T2S+, Gepco M1042, Mogami W2549, Redco LO-Z1. [Redco, Full Compass.]

- **Piezo discs.** These are also called benders. Only one is needed, but I suggest getting several, as these can be a little tricky to work with. Avoid the piezos at Radio Shack, because they already have the wires soldered on and are sealed inside a plastic casing. [Item 668-1017-ND from DigiKey, or any piezo disc 15mm diameter or larger from Projects Unlimited.]

Sources

The companies listed here are the ones I use for my electronics work. Except for Bracken Creek Audio, I have no relationship with any of them except as a satisfied customer. There are many other sources for these materials. These are the ones I have good experiences with.

- **Bracken Creek Audio** is my own company. I have put together a Supplies Kit of the materials you need to do the exercises, and as of this writing, several kits for audio gear are in the works. Bracken Creek Audio's website is www.hurdygurdy.com/brackencreek.

- **DigiKey** is my first stop for electronics components. Of the three big components suppliers, I have found their search engine and service to be the best. They also have a very useful smartphone app. If I can't find something at DigiKey, I look at Mouser and Newark.

- **Mouser** is the second of the big three suppliers, with a selection similar to DigiKey's. When I still used print catalogs, theirs was equal to DigiKey's, but over the years I have found their search engine and website less user friendly.

- **Newark** is the third of the big three, with perhaps the largest selection, a huge print catalog, and a decent but not stellar search engine. I've found obscure transistors at Newark that I couldn't find anywhere else.

- **Electronic Goldmine** is my source for the more offbeat items, including grab bags of components and practice pieces such as printed circuit boards. It's sort of an online electronics thrift store. If you find something you think you want, buy it right away, because they may run out next week and never get any more.

- **Redco** is my preferred source for cable and connectors. They sell cable by the foot, so you can get just what you need.

- **Full Compass** is another source for cable and connectors. I've had good service from them also.

- **Techni-Tool** is my preferred source for tools, with a great assortment, a very nice catalog, and decent prices.

- **Jameco** is a good source for breadboards, tools, parts, and some interesting kit projects.

1

Basic Soldering

Goal: When you have completed this chapter, you will have gained an appreciation and understanding of the reasons for using solder and the techniques of safely and effectively soldering and desoldering.

Objectives

- Choose the soldering station that is right for you.
- Practice safety while soldering, and be prepared with first aid if needed.
- Prepare the soldering iron and its station for use.
- Solder joints that are smooth, strong, and bright.
- Desolder joints cleanly with the pump and braid.
- Solder wires to a battery clip for use in later studies.

How Soldering Works

Solder provides a very solid connection that is not subject to vibration or most other physical trauma. It is fairly easily reversible, so components can be removed. Soldering is a basic skill in electronics. It is used for making cables, in assembling new equipment, and in making repairs. A soldered connection is called a *joint*, as shown in figure 1-1.

Old-fashioned solder is an alloy (mixture) of tin and lead, typically about 60% tin and 40% lead. It melts at a temperature of about 400°F. Solder used for electronics is available in many forms, but in most cases we use a thin wire form, as shown in figure 1-2.

Coating a surface with solder is called *tinning* because of the tin content of solder. Lead is poisonous, so you should always wash your hands after using solder. There is a common misconception that soldering generates lead fumes that you will breath. This is incorrect. Lead doesn't boil until it reaches a temperature of over 3000°F, which we're not even close to with a soldering iron at 750°F.

Because of the toxicity of lead, electronics with lead solder are being phased out. As of 2006, the European Union requires compliance with the Restriction of Hazardous Substances (RoHS) rules. These rules are broad reaching and have been felt in every aspect of electronics component manufacture. In particular,

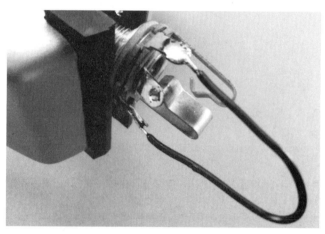

Figure 1-1.

Figure 1-2.

RoHS eliminates the use of all lead in electronics. While many electronics manufacturers and components makers are still using lead solder because of its ease of use and other qualities, it is fairly certain that almost all lead will be gone from the electronics world in the next few decades.

The big problem with eliminating lead from solder is the question of what to replace it with. The reason that lead was used in the first place is that it works so well and it is relatively inexpensive. It melts easily when mixed with tin, and it solidifies with a smooth finish. It also prevents the formation of "tin whiskers," which are fine hairs of tin that spontaneously grow from the soldered surface. Lead-free solders have higher melting points, and their joints look pitted and rough. They still work well, but it's harder to tell the difference between a good joint and a bad joint. It is also more difficult to keep the soldering iron tip tinned properly so that the solder flows easily. Lead-free solder is more expensive because of the costlier metals that are substituted for lead.

Flux

Solder for electronics contains tiny cores of *flux*. This flux chemically cleans the metal surfaces as the solder melts. Flux is corrosive, like an acid, when it is activated by heat. This is why you must melt the solder onto the joint, not on the iron tip. Without flux, most joints would fail (if they could be joined at all), because metals oxidize quickly, and the solder will not flow properly onto a dirty, oxidized metal surface. A badly corroded or oily surface will not let solder stick to it at all, regardless of the action of the flux. The flux in electronics solder is a rosin, made from pine pitch. Never use acid or salt fluxes for electronics. The joint may look all right initially, but it will fail quickly, and the circuit board and the component will probably be damaged as well. Similarly, never use a soldering gun on any kind of electronics. Besides being bulky and horribly imprecise, it generates a substantial magnetic field that can damage sensitive components.

Choosing a Soldering Iron or Station

Once you know what you are doing, it's possible to do very good work with an inexpensive soldering pencil, but it's harder to get started with one. If you are at all serious about doing electronics work, a good *soldering station* is one of your best investments. I used a succession of cheap irons for years before finally taking the plunge to a thermostatically controlled Weller station, and I wished I had bought one years earlier. It made a big difference in both what I was able to solder and the quality of my solder joints.

The Weller WES51 soldering station shown in figure 1-3 is an industry workhorse. The soldering tip is on a thermostat, meaning that it stays in a narrow temperature range. It doesn't overheat when it's sitting unused in the stand, and it doesn't lose temperature when it's being used for a long time on a

Figure 1-3.

heavy joint. The list price is pretty high, about $100 in 2014, but there's almost always a special sale somewhere on the Internet where you can pick one up for less.

The replacement tips for this model are relatively inexpensive and come in a variety of shapes and sizes. I find that a small screwdriver-shaped tip is best for most work, as shown figure 1-4.

Soldering Safety Precautions

Using a soldering iron is inherently dangerous: it's very hot, and it can expose you to toxic chemicals. A little common sense will keep you safe.

Figure 1-4.

- Always wear safety glasses. You can take your chances with your eyesight if you want to, but there are just too many opportunities for eye damage to risk it. Safety glasses are inexpensive, and many models fit over regular glasses if you wear them. There are also bifocal safety glasses if you need better close-up vision, as we tend to as we get older. If you are having trouble seeing through your safety glasses because they are scratched, throw them away and get another pair. Once again, a few dollars versus losing your eyesight is a pretty easy choice.

- Never touch the element or tip of the soldering iron! It is very hot (about 750°F) and will give you a nasty burn.

- Take great care to avoid touching the power cord with the tip of the iron. Ideally the iron should have a heat-resistant cord for extra protection, but no rubber or plastic insulation in the world will stand up to that temperature for very long.

- Always return the soldering iron to its stand when not in use. Never put it down on your workbench, not even for a moment!

- Work in a well-ventilated area. The smoke formed as you melt solder is from the flux burning off. It can be quite irritating. Avoid breathing it by keeping your head to the side or in front of your work, not immediately above it.

- Hot metal looks just like cold metal. Let the work piece cool before touching it.

- Wash your hands after using solder. As mentioned earlier, solder contains lead, which is a poisonous metal.

- To avoid accidentally ingesting lead, don't eat or drink at your workbench.

- Read the First Aid section so you will know what to do if you burn yourself.

First Aid for Burns

Most burns from soldering are likely to be minor, and treatment is simple.

- Immediately cool the affected area under gently running cold water. Keep the burn in the cold water for at least five minutes. If ice is available, this can be useful also. Cooling with ice helps to numb the burn while also reducing the tissue damage. Cooling the burn immediately is the key.

- Do not apply any creams or ointments. The burn will heal better without them. A dry dressing or bandage may be applied if you wish to protect the area from dirt.

Reduce the Risk of Burns

The best way to treat a burn is to never get one in the first place. Experience has taught me the habits to help avoid trouble.

- Always return your soldering iron to its stand immediately after use.
- Allow joints and components a minute or so to cool down before you touch them. At the risk of repeating myself, hot metal looks just like cold metal.
- Never touch the element or tip of a soldering iron unless you are certain it is cold. Unless you are replacing the tip, there really isn't any circumstance when you would need to touch it anyway, except maybe to see if it is cool enough to be put away in a drawer.
- If you think that something might still be too hot to touch, graze it with the backs of your fingers. If it really is too hot, your reflex reaction will be to pull away, and your fingers won't be pulled into the hot piece. (If you do get a burn, it will be on the backs of your fingers where there are fewer nerve ending. It's still a burn, but it's better than a burn on your fingertips.)

Preparing the Soldering Iron or Soldering Pencil

Before starting to work, the soldering station needs a little preparation. In particular, remember to check the tinning of the tip before starting work.

- Place the soldering iron or pencil in its stand. If you have an iron with a controller, plug the iron into the control unit. Turn on the controller, and set it to 750°F if it has a temperature control. If you have a soldering pencil, just plug it in. The iron will take perhaps a few minutes to reach its operating temperature. A soldering pencil will take longer than a temperature-controlled station.
- Dampen the sponge in the stand. The best way to do this is to lift it out of the stand and hold it under a cold-water tap for a moment, then squeeze to remove excess water. It should be damp, not dripping wet.
- Check the tinning at the tip of the iron by wiping the tip on the damp sponge and then melting some solder onto the tip. If the solder accumulates in a silvery liquid drip, then all is well. If the tip is not already silvery and it is hard to melt the solder, or the solder that does melt is immediately repelled by the tip, then the tip needs some attention. It may need to be cleaned with a special tip cleaner to replace the tinning on the surface, or just replaced.
- It's all right to leave the soldering iron turned on when you are working, but it's safer to turn it off when you walk away. Temperature-controlled stations heat up very quickly, and the tip will last longer if it doesn't sit in the stand for hours without being used.

Hands-On Practice

To get started soldering, you'll solder some wires into *solder lugs*, which are small, flat pieces of metal with holes through them. When you are starting out, a good practice piece is a 1/4" jack.

They are easily available, inexpensive, and pretty much indestructible. You'll also need some wire. I recommend solid core hookup wire, but you can use stranded wire also. The solid wire is easier to work with. An example of a completed soldering practice piece is shown in figure 1-5.

When you have completed several joints, you'll use two methods of *desoldering* the wires to get them disconnected and the lugs cleared. You can practice this as many times as you want to get comfortable with soldering. They say that Rome

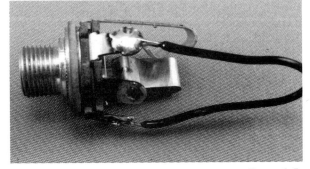

Figure 1-5.

wasn't built in a day, and similarly no one is born just knowing how to solder and desolder. It takes a little practice, just like any other skill. (If things go horribly wrong, you can just throw the sample work piece away and start fresh with a new one.)

Once you've practiced some soldering and desoldering with solder lugs, you can also make a soldered *battery clip* for using a 9V battery with your breadboard.

Materials

These are the materials you will need for these exercises.

- Safety glasses
- Soldering station or soldering iron
- Solder
- Insulated solid core wire (20, 22, or 24 gauge)
- 1/4" jack or similar jack with solder lugs on it for soldering practice
- Desoldering pump
- Desoldering braid
- Helping Hands jig
- Panavise or similar vise
- Needle nose pliers
- Side cutting wire cutters
- Wire strippers
- Battery clip for 9V battery (battery clip only)

Your First Soldering Example

Start by practicing on something that is easy to work with, like a 1/4" jack such as you would find in an effects pedal. Anything that has solder lugs with holes in it will do. You will be soldering some pieces of solid wire (20, 22, or 24 gauge) into these lugs.

Preparing the Wire for Soldering by Stripping and Tinning

• There are several varieties of wire strippers available, each with advantages and disadvantages. The most common look like flattened pliers with a set of holes, with half of the hole in each jaw. For most electronics work, a pair of these with holes ranging from 20 gauge (the smallest) to 12 gauge (the largest) will be more than adequate. There are also smaller adjustable strippers with a single hole and a sliding lock nut. These work in a pinch, but they are annoying because they need to be adjusted to change wire sizes. There are also self-setting strippers, which detect the wire size and remove the insulation with a lever action. I've found these useful for home and studio electrical wiring, but they are too imprecise for fine electronics work.

• Strip the end of the wire by placing the appropriate set of half holes of the stripper jaws around the insulated wire, closing the jaws firmly, and pulling the cut insulation off of the end of the wire. Use the largest set of stripper holes possible, so that the copper wire core doesn't get nicked. Solid wires nicked by strippers are at risk of breaking when flexed, and wires lost from stranded cable will reduce the effective diameter of the wire. Once you are experienced, you will aim to strip the shortest length of wire that is necessary to feed through the hole, so that the insulation will go right up to the solder joint. For now, just strip about 1/2", and trim it later if needed.

- Hold the wire in a Panavise, Helping Hands jig, or other holding device, with the stripped wire end available for soldering.
- Heat the wire by holding the soldering iron in contact with it. While still heating the wire, hold the tip of a piece of thin rosin core solder on the wire until it melts. Withdraw the solder and then the soldering iron. Coating the wire with a thin layer of solder in this way is called "tinning" the wire. This is shown in figure 1-6.

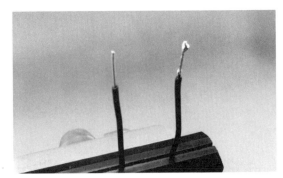

- The untinned wired is on the left, and the tinned wire is on the right. For the example piece, I've used a lot more solder than I usually use for tinning, so that you can see it. The usual goal of tinning is to just thinly coat the wire or lug.
- Let the wire cool for a moment (to prevent burning yourself), and remove it from the holding device.

Figure 1-6.

Preparing the Practice Piece

- Hold the connector in a vise or Helping Hands jig.
- Tin the connector surface in the same way you did for the wire. The goal of tinning is to provide a surface that already has the solder bonded to the metal, so that it will be easier to get the two surfaces to accept the molten solder when they are joined together. Tinning is not required, but it's a good idea because it helps keep the solder joints smooth, bright, and clean. Ultimately this results in better-quality solder joints that last longer and stand up to more abuse. In the first soldering photo below, you can see that the lug is bright with the solder from tinning.
- Trim the stripped end of your wire to about 1/4" long.
- Use needle nose pliers to bend the wire into a hook shape.
- Place this hook into the hole in the lug, and crimp it tight with the needle nose pliers. Ideally it will then stay in place without being held.
- If needed, hold the wire in position with the Helping Hands. The wire should be extending straight out from the lug.

Ready to Start Soldering

- Hold the soldering iron like a pen, near the base of the handle. Remember to never touch the hot element or tip.
- Wipe the tip of the iron on the damp sponge with a rolling motion. This will clean the tip and make it silvery.
- Touch the soldering iron onto the joint to be made. Make sure it touches both of the parts to be soldered, and hold it in as much contact as possible, as shown in figure 1-7.

Figure 1-7.

- Hold the tip there for a few seconds and feed a little solder onto the joint. It should flow smoothly onto the parts. Apply the solder to the joint, not the iron! Solder flows toward heat. The iron will always be hotter than the parts being joined, so solder applied to the iron will stay there rather than go down onto the joint where you want it.

- Remove the solder, then the iron, while keeping the joint still. Allow the joint a few seconds to cool before you move anything. The solder should be smooth and bright, and it should fill all of the gaps around the wire where it goes through the lug, as shown in figure 1-8.

- Avoid touching the work until it cools. Hot metal looks just like cold metal!

- Inspect the joint closely. It should look shiny and smooth, with the solder flowing over both pieces being soldered. If it doesn't look smooth, you will need to reheat it and feed in a little more solder. This time ensure that both the wire and the connector are heated fully before applying solder. Dull or clumpy solder is an indicator of a cold joint that will not stand up or may not even be really bonded at

Figure 1-8.

all. If the joint came out well, congratulations! If it didn't, desolder it, clean the connector, and try again. In either case, practice this joint several times before moving on to making the battery clip. You'll need to make several joints to practice desoldering.

Desoldering

At some stage you will need to desolder a joint to remove or reposition a wire or component. There are two ways to remove the solder: with a *desoldering pump* and with *desoldering braid*. Try them both on your practice joints. Some people prefer the pump; others prefer the braid. I use them both, depending on the situation.

Using a Desoldering Pump

The desoldering pump is a hand-powered suction pump that operates on a spring-loaded plunger. It's typically just called a *solder sucker*. It has the advantage of durability. It is sometimes harder to handle and less precise than the braid.

- Prepare the pump by pushing the spring-loaded plunger down until it locks.

- Press the button to release the plunger. Try the lock-and-release cycle several times until you get a feel for how the pump works and how much it bounces around when released.

- Prepare the pump again, this time for removing solder.

- Apply the tip of your soldering iron to the joint, holding the tip of the desoldering pump close by, just out of reach of the iron tip. This is shown in figure 1-9.

- Wait a few seconds for the solder to melt.

Figure 1-9.

- Press the button on the pump to release the plunger and suck the molten solder into the tool. The release and movement of the plunger happens very quickly, and it can be a little startling the first few times. This rebound makes it a little challenging to use a solder sucker precisely until you get accustomed to it.

- Repeat if necessary to remove as much solder as possible.

- Occasionally the pump will need emptying. This is accomplished by unscrewing the nozzle and tapping out the globs of dead solder. Don't bother trying to reuse this solder for anything, as it no longer has any flux in it and it won't stick to anything.

- After removing most of the solder from the joint, you may be able to remove the wire right away (allow a few seconds for it to cool). If the joint will not come apart easily, apply your soldering iron to melt the remaining traces of solder while pulling the joint apart, taking care to avoid burning yourself. When you are done, the solder lug should be bright and clean, with just the thinnest coating of solder remaining.

Using Desoldering Braid

Desoldering braid is a ribbon of fine wires braided together and saturated with rosin flux. When the braid is heated, capillary action from the high surface area pulls melted solder up into it, just as a wick does. It's a consumable, in that once it soaks up the solder, it isn't useful any more. It often gives a cleaner result than the solder sucker, but it also leaves a residue of flux.

Figure 1-10.

- When using the braid, leave it in its dispenser roll. If you pull out a piece and cut it off, the braid will heat in your hand and become too hot to hold.

- Place the end of the braid on top of the joint.

- Apply the tip of your soldering iron to the braid, in full contact, as shown in figure 1-10.

- As the solder melts, most of it will flow onto the braid, away from the joint. If there is a lot of solder, you will need to move the braid along, passing it over the joint as it wicks the solder up into it. This is shown in figure 1-11.

- Remove the soldering iron and the braid.

- The braid will still be hot. Let it cool. Once again, remember that hot metal and cold metal look exactly the same.

- Cut off and discard the end of the braid coated with solder.

Figure 1-11.

- With most of the solder removed by the braid, the wire should come free easily. If not, reapply the braid as before to remove more solder.

Making the Battery Clip for Your Breadboard

You will need a battery clip for your breadboard work in the upcoming chapters. Here you will solder pieces of solid wire onto the stranded wires of the battery clip so that it's easier to insert the wires into the holes of the breadboard. This will allow you to use the battery clip over and over. Without these solid wire ends, the stranded wires of the battery clip must be inserted and left in place in the breadboard.

- The starting materials are shown in figure 1-12.

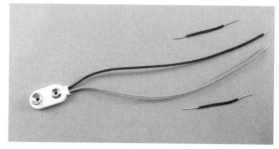

Figure 1-12.

- Cut two pieces of solid wire about 2" long.

- Strip about 1/2" from each end of the two wires.

- Cut off the tinned ends of the battery clip, and strip them about 1/2". Twist together one end of a solid wire with the stripped end of the battery clip's black wire. This is tricky because the stranded wire is thinner and much more flexible than the solid wire, so the stranded wire is being wrapped around the solid wire.

- Hold the two wires in the Helping Hands jig or Panavise.

- Solder together the two wires by heating them with the soldering iron, as shown in figure 1-13. Notice that the soldering iron tip is angled to bring it into as much contact as possible with the wires. This allows the heat from the iron to be transferred faster and heat up the wires more consistently.

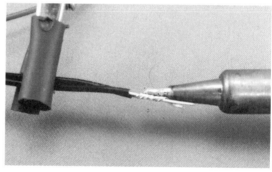

Figure 1-13

- Once the wires are hot, add solder directly to the wires so that it melts over them and holds them together, as shown in figure 1-14.

- Repeat the procedure for the second piece of solid wire and the red wire.

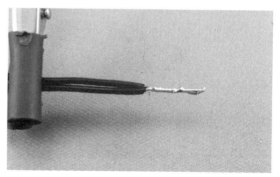

Figure 1-14.

- The completed battery clip is shown in figure 1-15.

- Store this clip with your materials for later use with your breadboard.

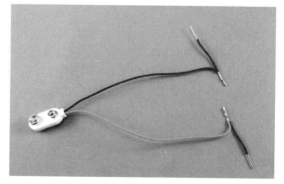

Figure 1-15.

Voltage, Current, Resistance, and Building a Circuit on the Breadboard

2

Goal: When you have completed this chapter, you will have gained an understanding of the relationship between the core electronics concepts of voltage, current, and resistance.

Objectives

- Identify electromotive potential as the driving force in electronics.
- Define the key concepts of voltage, current, and resistance.
- Recognize the equation symbols and units for these quantities.
- Recognize the circuit diagram symbols for battery, resistor, and LED.
- Use a circuit diagram to build a circuit on a solderless breadboard.

Voltage

Voltage is the basis of everything in electronics. As we will see, voltage is the driving force behind all of the other concepts of what electricity is and how it works.

The Structure of the Atom

To understand electricity, we need just a little background in chemistry and physics. All matter is made up of elements. Elements cannot be reduced to simpler substances by normal chemical means. Examples of elements are oxygen, hydrogen, copper, iron, carbon, and so on. An *atom* is a single particle of an element.

Atoms are made up of combinations of three even smaller particles, called *protons*, *neutrons*, and *electrons*. Protons have a positive charge and considerable mass. Neutrons have the same mass but no charge. Together protons and neutrons form the nucleus, the tight, positively charged core of the atom. A cloud of electrons surrounds the nucleus. These electrons have a negative charge and essentially no mass, as shown in figure 2-1.

The universe likes to be a balanced place, so the number of negative electrons equals the number of positive protons. In this way, the atom's overall charge is zero.

We are interested in the electrons at the outer edges of the atoms of some elements, particularly copper. These electrons can be stripped away and pulled off somewhere else. When we do this, two little parts of the universe are now unbalanced. The atom is missing an electron, so it's positively charged, and it wants the electron back. The electron, out away from the atom, is negatively charged.

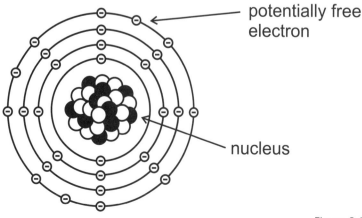

potentially free electron

nucleus

Figure 2-1.

It wants to get back "home" to its nucleus, or to any other positively charged place where its negative charge will be balanced out. The electron's desire to get back "home" is what makes electricity work, and makes all of the electronic marvels in our modern world possible. The electron is the part that moves, as it's so much lighter and more mobile than rest of the atom.

To get the electron away from its home atom, we put a certain amount of work into pulling it away. The electron now has that much energy, and it is this energy that does work for us while the electron is making its way back home. We can examine a group of electrons and determine how much work they are capable of doing. This quantity is formally called *electromotive potential*. In an equation, we use the letter **E** to symbolize it. It is measured in *volts*, which are symbolized by the letter **V**. Electromotive potential is more commonly called *voltage*. The volt is named after the Italian Count Alessandro Volta, who invented the electric battery around the year 1800.

Visualizing Voltage

Voltage is the first of the three major concepts that are the basis for all electronics, so we need to have a good image in our minds of what it is and how it works. Voltage is similar to the potential energy that an object has when we raise it up in the air. For example, if we have a book on a table, and we lift it up a few inches or centimeters, the book now has potential energy. If we let it go, it will fall back to the table. If your hand is underneath, it will hurt a bit. If we lift it higher, when we let it go, the force will hurt your hand more. (I'm not recommending this, of course.) The point of this silly example is that we don't need to actually drop the book to find out how hard it will hit. All we need to do is look at how high the book has been raised, and this will give us an idea of how much it will hurt if it drops. The book's distance above the table is a rough measure of the potential damage it will do to your hand.

In the same way, we can measure the voltage of a group of electrons. We choose a reference point. Next we measure just how hard the electrons at another point are wanting to get to the reference point. The farther they are from home, the more work they can do as they travel home and the more desperately they want to get there. Physical distance doesn't matter here. The more important factor is how much work went into removing the electron from its home atom.

We can think of a battery as a container with two compartments: one full of free electrons, the other full of the atoms that are missing those electrons. Given an opportunity, the electrons will move along an available pathway to get back to their home atoms. A battery is actually a chemical reaction that continuously separates electrons from their atoms, but for the purposes of thinking about how electricity works, it's easier to think of the electrons as already being separated.

A useful model for considering electricity is to think of it as being like water, as shown in figure 2-2. If we pump some water out of the ocean into a tank on a dock, the tank is like a battery. The water in the tank wants to get back to ocean, but we're restraining it in the tank. If we open a pipe from the tank back down to the ocean, the water will flow out, and work can be done by that water flow.

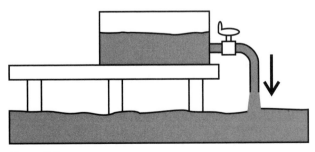

Just as the water in the tank could do work as it flows down the pipe to the ocean, electromotive potential is stored electrons impatiently waiting to find a way back to their home. If they are at a high potential, they can do a lot of work on their way back. If they are at a low potential, they can only do a little work.

An important point about potential is that when measuring it, work is not necessarily being done. If we have a tank of water at the top of

Figure 2-2.

a 200-foot tower and another at just 10 feet above sea level, we would say that the water in the tower is at a much higher potential than the other, as shown in figure 2-3. The water does not need to be flowing for us to measure the height of the water and calculate how much more work the water in the tower tank can do. We just have to see that there is some water in the tank and how high off the ground the tank is.

Similarly, we can compare a 1.5V battery with a 9V battery. The collection of charged particles in the 1.5V battery have a certain distance that they can fall to get from the (−) terminal to the (+) terminal. While falling, there's a certain amount of work they can do. The particles in a 9V battery have much farther to fall, so they can do a lot more work.

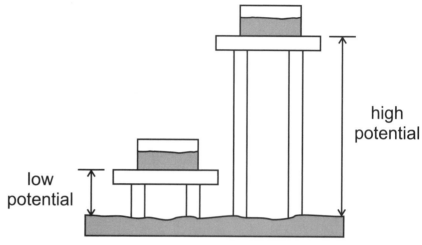

Figure 2-3.

Just as with measuring the water in the low and high tanks, when we measure potential, we're measuring how far the particles can fall, not how far they are falling. They don't actually have to be falling to measure how far they can fall. An example is when we measure the voltage (or potential) of a 9V battery sitting by itself. The battery isn't part of a system where work is occurring, but we can still measure how much work it can do. We can represent the potential held by these 9V and 1.5V batteries graphically, as shown in figure 2-4. This looks very similar to the earlier figure of the two water tanks.

With a digital multimeter (DMM) set up as a voltmeter, we can measure the voltage across the terminals of a 9V battery. If we attach the black probe to the (−) terminal and the red probe to the (+) terminal, we'll measure a potential of about 9V. If we reverse the probes, putting the black probe on the (+) terminal, we'll measure a potential of −9V.

This demonstrates the importance of deciding where to measure voltage from. The black probe of the DMM can be moved around, and the potential of other points is then measured from that point. Both of these potentials are measured from a common reference, the zero line. In electronics, we call a zero reference point the *ground*. Many measurements in electronics are made relative to the ground point.

There's another important point here. Regardless of where we're measuring from, when we measure voltage, we are measuring the potential between two points. Voltage does not move from place to place. It exists at one point when measured relative to another.

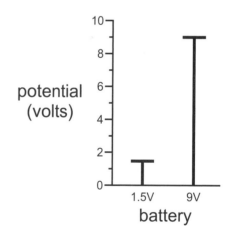

Figure 2-4.

Choosing Between E and V

In an equation, the letter E is the standard symbol for the quantity of electromotive potential, and it's the one we'll be using in this book. Other books use the letter V, because electromotive potential is commonly called voltage. My preference is to use E for the quantity and V for the unit of volts. This keeps it clear when we are talking about the quantity of potential and when we are using the unit with a number.

The Battery

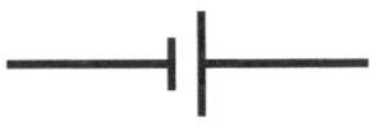

The circuit diagram symbol for a battery cell is shown in figure 2-5. The pair of vertical lines represents a chemical cell, which is a pair of metal plates that perform a chemical reaction that generates about 1.5V. The shorter line represents the electron source that is negatively charged, and the longer line represents the positively charged home atoms that they want to get back to.

Figure 2-5.

If we stack several cells together, we get a battery. Several of the variants of the circuit diagram symbol are shown in figure 2-6. The taller line at the right side of the symbol indicates the positive pole of the battery, regardless of whether the (+) and (−) signs are included. All four of the symbols shown are valid for a 9V battery. The (+) and (−) signs shown in the first symbol are usually omitted, though it's not technically wrong to put them in. A purist would say that there should be six sets of lines, one for each 1.5V cell, but it's all right to just draw several sets of lines and write the voltage nearby, as shown in the last symbol.

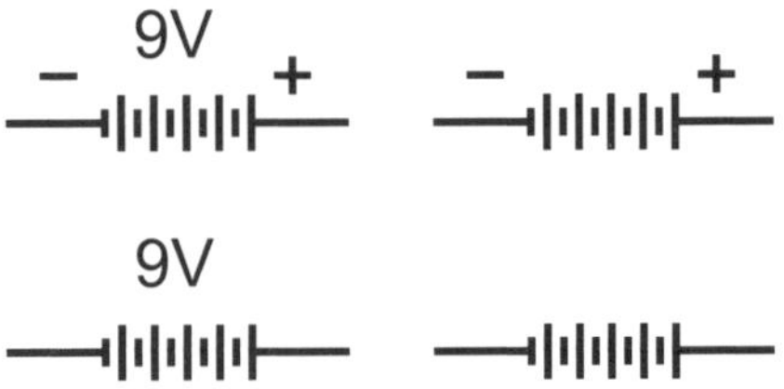

Figure 2-6.

The meter for measuring electromotive potential or voltage is the voltmeter, which has the circuit diagram symbol shown in figure 2-7. It should be noted that the symbol indicates not just the meter, but also the probes or wires that are used to connect it to the circuit to make the measurement. A digital multimeter (DMM) has several functions, one of which is to act as a voltmeter.

Figure 2-7.

Current

When we have a voltage source like a battery, essentially we have a collection of electrons wanting to get back to their homes. If we provide a path for them to follow from their source to their home atoms, they will move along that path as fast and furiously as they can. The movement of electrons from one place to another is called *current*. In an equation, this quantity is symbolized by the capital letter **I**. This seems a little strange, because there is no "i" in the word *current*. The symbol was established by the French physicist André-Marie Ampère, who described the "intensité de courant" (French for "current intensity") in his writings on electromagnetism. Hence **I** stands for current intensity, which we just shorten to current. Monsieur Ampère also gave his name to the unit of current, the *ampere*, which we often shorten to *amp*. The abbreviation or symbol for the unit is a capital **A**. Any wire or other material that has current flowing in it is called a *conductor*. Current is the second of our essential three concepts of electronics.

The last few years have seen an increasing use of the word *amperage* in place of the word *current*. This is similar to *voltage* being used for electromotive potential. While its use is regrettably becoming more common, it's still a really ugly word. The word *current* is correct, while the other word is amateur. It's gotten as far as being printed on appliance labels, but it's still wrong as far as I'm concerned.

The meter for measuring current is the *ammeter*, which has the symbol shown in figure 2-8. Just as for the voltmeter, the symbol indicates not just the meter, but also the probes or wires that are used to connect it to the circuit to make the measurement. These two connections are indicated by the two lines at the sides.

Figure 2-8.

Resistance

The restriction of the flow of current is called *resistance*. In an equation, resistance is represented by the letter **R**. The unit of measurement for resistance is the *ohm*, named after the German researcher Georg Simon Ohm, who studied electricity during the early 1800s, soon after Volta's development of the battery. The symbol for the ohm is the Ω, which is the capital letter omega from the Greek alphabet

A resistor is an electronic component that has a certain amount of resistance. Figure 2-9 shows the circuit symbols for a resistor. The old symbol shows that the current needs to go on a little zigzag course through the resistor, which restricts how much flows through. The new symbol is just a box. It's easier to draw, but I favor the old symbol, because it conveys not just where the resistor is, but also what it does.

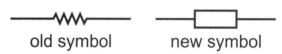

old symbol   new symbol

Figure 2-9.

In circuit diagrams, a resistor is labeled with an **R**: R1, R2, R3, and so on. This labeling is also seen on circuit boards. Some typical resistors are shown in figure 2-10. The standard form of a resistor is a small cylinder with a wire coming out of each end. This wire of the component that gets soldered into a circuit board is called a *lead*, which rhymes with *feed*. (This is different from the word *lead* in solder, which rhymes with *head*).

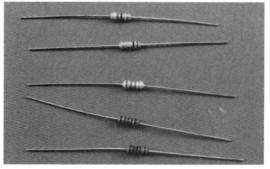

Figure 2-10.

We can figure out just how much resistance the component has from its identifying marks. Resistors are manufactured in a dizzying variety of shapes and sizes, but most of them are cylinders marked with four or five colored bands. Reading the color code of the resistor will be discussed later in the book. Some of the other forms of the resistor are shown in figure 2-11. The tiny resistors shown toward the bottom right are Surface Mount Technology (SMT) resistors. These are used in new equipment in place of the older through-hole resistors with the two leads soldered into the circuit board. You will also see them called Surface Mounted Devices (SMDs).

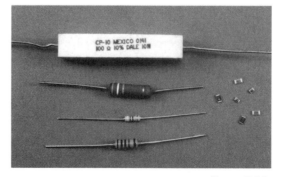

Figure 2-11.

The *ohmmeter* is used for measuring resistance, and its circuit diagram symbol is shown in figure 2-12. Again, the symbol indicates not just the meter, but also its two connections to the circuit.

Resistance is the third of our three essential electronics concepts. With voltage and current, it forms the basis of all electronics. The key to understanding resistance is to think of it as having the ability to restrict or reduce current. Voltage is the pressure to move electrons around, current is the movement of

Figure 2-12.

those electrons, and resistance is what holds them back from all rushing through at once. For example, a piece of wire has essentially no resistance. If we use a piece of wire to connect one battery terminal to the other, the battery drains very quickly, because there is hardly anything holding back the flow of electrons. We don't want to use up a battery this way, as it is just wasteful. We'd rather use the current to do some useful work for us, like powering a little pair of headphones or making a radio connection to a cellphone tower. To hold the flow of current down to a trickle, we use resistance.

The Light-Emitting Diode (LED)

We have now seen several examples of circuit diagram symbols. To explore circuit building and circuit diagrams, we need one more symbol. This one is for a light-emitting diode, which we usually just call an *LED* (pronounced with the three letters spelled out: "L-E-D"). The LED is the ubiquitous little light used as an indicator in electronics. In the hands-on exercises, we will use them to show us whether or not there is current flowing. In a circuit diagram, an LED is usually labeled with a **D**: D1, D2, and so on. Figure 2-13 shows the circuit diagram symbol for an LED. Figure 2-14 shows a typical LED.

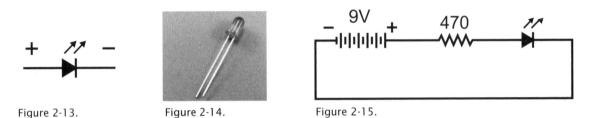

Figure 2-13.   Figure 2-14.   Figure 2-15.

We'll study light-emitting diodes and their function more in a later chapter. For now we just need to know the symbol and how it relates to the real object. Like the battery, the LED has a specific orientation, or polarity. It only functions when electric current enters from the positive side and flows to the negative side. This direction of current flow is shown by the direction of the large triangle in the symbol.

To know which way to orient the LED when we are working with it, notice that the lead on the positive side of an LED is longer. We can also recognize the negative side in two ways: the lead is shorter, and there is often a flattened spot on the outside of the plastic case.

Circuits and Circuit Diagrams

A circuit is a path that electrons can follow to get back to their home atoms. Any time current is flowing from one place to another, the arrangement of components that it is flowing through is called a *closed circuit*. If we disconnect a component and break the path, this circuit is now an *open circuit*.

Figure 2-15 shows a circuit diagram, usually called a *schematic*. These terms mean the same thing and get used interchangeably. A circuit diagram is a standardized way of drawing out how components are connected to each other. It is made up of the symbols for components and lines representing conductors. The circuit diagram shown here is the first circuit you will be building on your breadboard in the hands-on exercise.

Each component has a specific symbol associated with it. We've already seen the symbols for the battery, the resistor, and the LED. More symbols are introduced throughout the book. Some symbols have a value listed or specified for the component. In figure 2-15 we see that the resistor is specified to be 470Ω. We don't need to include the Ω symbol, because it's understood that a resistor will be measured in ohms.

With only a circuit diagram and a parts list, you can create a working circuit. A circuit diagram is essentially a recipe for a circuit that performs a particular function, such as a microphone preamp, a guitar amplifier, a mixer, a compressor, an analog-to-digital convertor, and so on. The circuit diagram for a particular piece of gear is very helpful for troubleshooting and repairing it.

Current in a circuit diagram generally flows from upper left to lower right. This is a very general rule, so there will be exceptions. This rule also applies to audio signals: in general, inputs will be on the left side of the diagram, and outputs will be on the right side.

With practice and experience, you will be able to determine what a circuit is designed to do just by looking at the diagram. Don't expect this knowledge to come immediately. Reading a circuit diagram is similar to reading text or reading music. In learning any of these reading skills, it takes time to recognize groups of symbols as units instead of looking at them as individual letters or notes.

The Solderless Breadboard

Now that you can know the symbols for a battery, a resistor, and an LED, you can assemble the circuit shown in the previous figure. The best place to do this is on a *solderless breadboard*.

Electronic experimenters often find it useful to put together components in a temporary setup to test the circuit and see if it works before committing the time and effort involved in soldering them all to a circuit board. In the old days, when radio was a new thing and radio enthusiasts built radio sets at home, they would often lay the components out on a flat surface, pound in some copper nails, and use these as contact points to connect up the components before mounting them permanently. The flat surface most easily available was often a breadboard from the kitchen, and the practice of dry assembly became known as breadboarding. (What the ladies of the house had to say about the breadboard being used for this purpose was not recorded.) Long after we have moved on to more sophisticated methods of dry assembly, we still call it breadboarding.

The modern solderless breadboard shown in figure 2-16 is a tool for assembling components together and testing circuits easily. There are many models of breadboard, but they all work the same way. Each of the small holes in the surface of the breadboard has a little metal clip underneath it. When a wire is inserted into the hole, as shown in figure 2-17, the clip grips it and

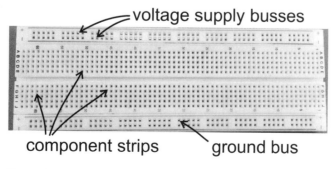

Figure 2-16.

makes a connection. The clips are arranged in rows that are connected to a common strip of metal. To connect two or more components together, place the tips of their wires in holes in the same component strip. We can think of these strips as being parts of a miniature patch bay.

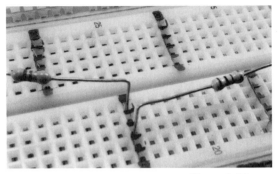

Figure 2-17.

All of the leads (wires) that we plug into a single component strip get connected to each other. Most of the holes on the breadboard are for the short component strips, which run up and down. In addition, there are four long strips along the edges of the board, two at the top and two at the bottom. We typically use the one at the top as a distributor for the voltage supplies and the one at the bottom as the ground point or ground bus.

If you are new to the breadboard, I recommend using a permanent felt pen such as a Sharpie to make some marks on your breadboard to remind you in which directions the strips run. This marking is shown in figure 2-18. I've used different colors for the (+) voltage bus at the top, the ground bus at the bottom, and the component strips. You may find that these markings help you get started on the breadboard. I marked just a few of the component strips, and we will use these by preference in the upcoming exercises. The marking will help remind you that the component strips do not connect to the buses at the top and bottom, and they make no connection across the gap in the center.

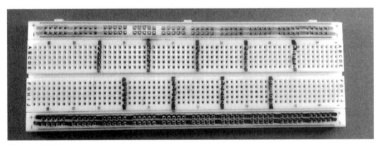

Figure 2-18.

For many of these exercises, you need a voltage source. The easiest way to attach a 9V battery to the board is with a battery clip. You have two options here. If you have access to soldering equipment, you can solder a piece of solid wire to the ends of the stranded battery clip wires. A battery clip with this modification is shown plugged into the breadboard in figure 2-19. If you don't have soldering capabilities, there is another alternative, which is explained in the directions of the hands-on exercise.

If you really start experimenting with building electronics on a breadboard, you may want to consider getting a DC bench supply. This piece of equipment usually has two voltage supplies that can be turned up or down independently. The bench supply never runs out, and you can choose any voltage instead of being stuck with 9V. It provides a regulated voltage supply, meaning that it keeps the voltage constant regardless

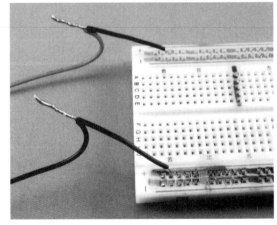

Figure 2-19.

of how much of a load you put on it. A typical bench supply is shown in figure 2-20.

Hands-On Practice

For the first few exercises in the book, I am assuming that you have never seen any of these materials before, so the instructions are very detailed. Over time and as you gain experience, I will make the instructions shorter and more general. You can always refer back to these earlier exercises for the more detailed directions.

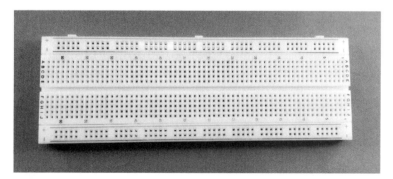

Figure 2-20.

Materials

These are the materials you will need for these exercises.

- Breadboard
- One or more Sharpie permanent markers (red, black, and blue are preferred, but any colors or just a single color will work)
- 9V battery and battery clip (with soldered wires if available)
- 100Ω resistor (banded brown–black–brown–gold)
- 470Ω resistor (banded yellow–violet–brown–gold)
- 1.0kΩ resistor (banded brown–black–red–gold)
- Red LEDs
- Green LEDs
- Insulated solid wire (20, 22, or 24 gauge)
- Needle nose pliers
- Wire strippers
- Rubber band or electrical tape (optional)

Preparing the Breadboard

It's easier to use the breadboard and visualize the connections being made by rows of metal clips if you add two jumper wires and mark some of the lines of holes on the breadboard surface.

• Turn the board as shown in figure 2-21, with the red line at the very top. The numbers and letters at the sides are now right side up.

Figure 2-21.

• Use a Sharpie to mark five vertical lines of five holes in the top half of the middle of the breadboard. The lines marked in the example are at #60, #50, #40, #30, and #20. If you have more than one color, blue is the preferred one here. The Sharpie dries slowly on plastic, so leave it alone for a while if you don't want it to smear.

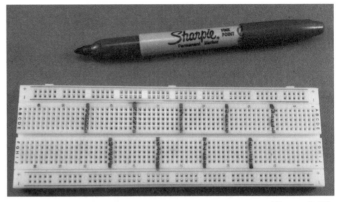

Figure 2-22.

• Mark four sets of lines in the lower half of the center. These marks should be in between the lines you marked in the upper half, as shown in figure 2-22. In the example, the lines marked are at #55, #45, #35, and #25.

• There are two lines of holes across the top. Marking them is optional. If you have another Sharpie (preferably red), use it to draw two lines across all of these holes, as shown in figure 2-23. Red is the preferred color because it often indicates positive voltage in electronics.

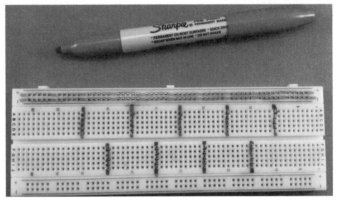

Figure 2-23.

• There are two lines of holes across the bottom. Marking them is optional. Use a black Sharpie to draw a line across all of these holes, as shown in figure 2-24. Black is preferred for this because it often indicates negative voltage or ground in electronics. Again, if you don't have black, any color will do.

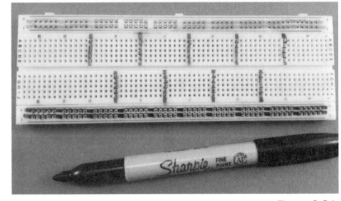

Figure 2-24.

- Strip about 1/2" off of each end of a piece of solid wire, as shown in figure 2-25. Cut off these stripped ends.

- Bend the bare wires into hoops to fit into adjacent holes in the breadboard. Needle nose pliers are helpful here.

- Place a hoop vertically to connect the two lines of holes along the bottom of the breadboard, as shown in figure 2-26. The hoop can be wherever you want it to be along the two lines, but at the end is convenient.

- Place another hoop to connect the two lines of holes along the top of the breadboard.

- These wire hoops serve two purposes. They make it easier to use the breadboard, because the entire top two rows of the breadboard are the (+) voltage supply, and the entire bottom two rows are the ground. This makes it less likely that you will miss when connecting something to either the voltage supply or the ground. These hoops are also good places to connect your digital multimeter (DMM) when taking measurements.

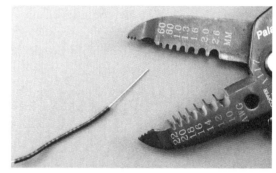

Figure 2-25.

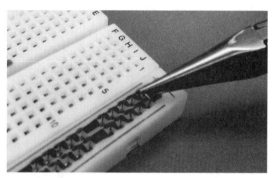

Figure 2-26.

Using a Battery Clip with Solid Wires

If you have a battery clip from the soldering exercise, use it here. If your clip doesn't have soldered wires, skip to the instructions below.

- Turn the board upright with the (+) bus at the top.

- Insert the stripped end connected to the red wire into the (+) voltage supply bus. If you are right-handed, it's more convenient to make this connection on the left side of the breadboard.

- Insert the stripped end connected to the black wire into the ground bus, as shown in figure 2-27.

- Plug a 9V battery into the battery clip.

Using an Unmodified Battery Clip

If you have an off-the-shelf unmodified battery clip, use it here.

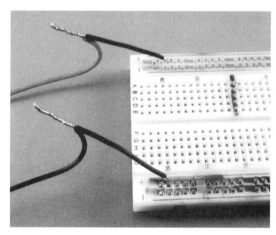

Figure 2-27

- The stranded wire of the battery clip is too flexible to be repeatedly pushed into the holes of the breadboard, so our goal is to install it more permanently. The ends of the stranded wires are usually twisted and tinned with solder for easier soldering, so they are a little stiffer than bare stranded wire.

- Turn the board upright with the (+) bus at the top.

- Insert the stripped and tinned end of the red wire into the (+) voltage supply bus, about ten holes in from the end. If you are right-handed, it's more convenient to make this connection on the left side of the breadboard. You'll need to grab the end firmly to push it in without it fraying or bending.

- Repeat this with the black battery clip wire into a hole in the ground bus. These inserted wires are shown in figure 2-28. You can leave the wires like this, but even the mild stress of moving the breadboard and battery around will soon make the connections unreliable.

Figure 2-28.

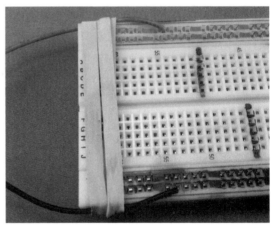

Figure 2-29.

- It's better to add some strain relief. You can wrap the end of the breadboard with electrical tape to secure the wires to the breadboard, or you can use a rubber band. (The rubber bands used to hold together bunches of asparagus, broccoli, or carrots work very nicely.) This is shown in figure 2-29.

Building the First Circuit

The first circuit will familiarize you with the node-by-node procedure of circuit building.

- The circuit diagram for the first circuit is shown in figure 2-30.

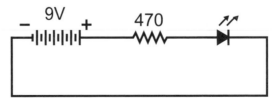

Figure 2-30.

- Neatness counts! Certainly you can put together a circuit on a breadboard with all the wires running everywhere, and most of the time it will still work. It's much easier to put together a circuit and follow its function when the components are arranged neatly. When a resistor or diode is used, it's good practice to bend the leads to 90° with a pair of pliers. You are aiming to make the leads into a U shape with fairly sharp corners. This is shown in figure 2-31.

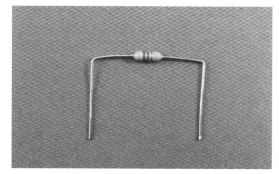

Figure 2-31.

- The voltage supply is a low enough voltage that it is safe to build these circuits with the supply plugged in at all times.

- The key to building a circuit is to build it one connection at a time. At every connection between two or more components, look at how many things are connected at that point. Plug each of them into a marked line or column on the breadboard so that they are connected together.

- For each component that you plug in, don't worry about where the other end goes yet. Just concentrate on making the connection between the first components.

- Start at the left of the circuit diagram, with the power supply providing a positive (+) voltage at the red lead of the clip. By plugging in the battery clip, you have already brought this voltage as far as the (+) voltage supply bus on the breadboard.

- The next component is the 470Ω resistor. Check the color code of the resistor to make sure it's the right value. The color bands for 470Ω are yellow–green–brown–gold.

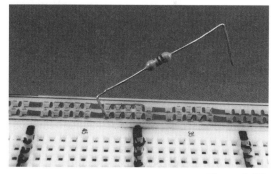

Figure 2-32.

- One of the resistor's leads gets inserted into the (+) voltage supply bus. You have now completed the first connection, between the supply and the resistor. We also call this connection a *node*, which is any connection between two or more components. This is shown in figure 2-32.

- The next node connects the second lead of the resistor to the LED. The resistor's lead gets inserted into one of the marked component strips on the board.

- The last component is the LED (red or green). As described above, the longer lead is the one that is connected to or closest to the (+) source (in this case, by way of the resistor). Identify which is the long lead.

- Insert the long lead into one of the other holes of the marked component strip that already has the resistor plugged into it. At the same time, insert the short lead into a hole in the unmarked component strip right next to the long lead. Remember that the component strips run up and down. You've now completed the second node. This is shown in figure 2-33.

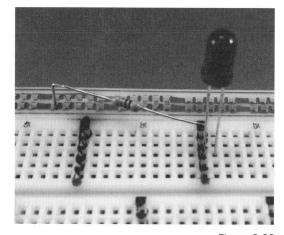

Figure 2-33.

- The last node is the connection between the LED and the ground. You'll use a piece of wire to make the connection, rather than stretching out the LED's leads. The short lead of the LED is already inserted into the unmarked component strip to the side of the previous one.

- Insert one end of a jumper wire (a short piece of wire with the ends stripped) into the LED's component strip (the one with the LED's short wire).

- Insert the other end of the jumper wire into the ground bus (the one that the black battery lead is plugged into).

- You've now completed the third node. The LED should light up. This is shown in figure 2-34. If the LED is not lit, check the troubleshooting suggestions that follow.

This exercise shows you that you can start with a circuit diagram and create that circuit on the breadboard.

Figure 2-34.

Troubleshooting

If the LED doesn't light up, here are some things to try.

- Check that each wire is firmly inserted in its socket.

- Check that the connections are actually being made between the components: the (+) voltage bus runs side to side, the component strips run up and down, and the ground bus runs side to side. There is no direct connection between the (+) bus and the component strips below. For a component to be connected to the (+) supply, it must be inserted into the (+) bus.

- There is no connection between the upper component strip (five holes) and the lower component strip (five holes). For two components to be connected, they must be in the same set of five holes.

- Make sure that the LED is plugged into two different component strips. If both leads are plugged into the same strip, one above the other, the LED will not have current flowing through it.

- Check that the LED is oriented correctly, with the long wire connected to the resistor.

- The LED may be burned out. Try a different one.

Visualizing Current

It is instructive to visualize or imagine the current flowing in the circuit. When the circuit is completed and the LED is lit, we know that electrons are moving through the wires and components, and doing work for us by creating light. It may seem a silly part of the exercise to do something so nonscientific as to use visualization, but I recommend doing it anyway. We can't see the current flowing anywhere in our modern world, but it's all around us, and we know it is there because we see the work it does. Visualizing the current flow helps make it more real for us. This will help us later when we are measuring voltage and current and resistance, and when we are analyzing what a particular circuit does.

Take a look at the circuit. You know that there is current flowing in the circuit because the LED is lit up. Visualize or imagine where the current is running:

– from the (+) terminal of the battery
– through the red wire of the battery clip
– into the (+) voltage bus on the breadboard
– through the metal strip of the supply bus
– into the wire of the resistor
– through the resistor
– through the resistor's other wire
– through the metal of the connection strip
– through the (+) wire of the LED
– through the LED (and producing light)
– through the (–) wire of the LED
– through the metal of the connection strip
– through the jumper wire
– through the metal strip of the ground bus
– through the black wire of the battery clip
– into the (–) terminal of the battery

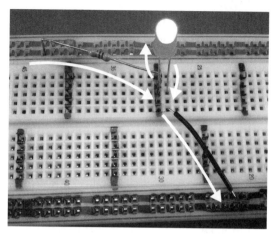

Figure 2-35.

This current flow is shown in figure 2-35. Notice that the current in the circuit goes through each of the components one after the other. This exercise shows that you can visualize how current flows from the positive terminal of the battery to the negative terminal. In later labs with the DMM, we will explore more precise ways of seeing this current flow and the voltage that pushes or pulls it along.

Building the Second Circuit

In this circuit, we will use two resistors instead of just one.

• Remove the resistor, the LED, and the jumper wire from the breadboard.

• The diagram of the second circuit is shown in figure 2-36.

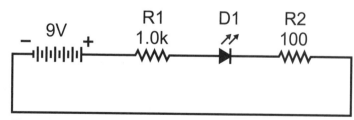

Figure 2-36.

• Assemble this circuit on your breadboard. Follow the node-by-node procedure of assembling a circuit, just as you did for the first circuit. You can do this by following the circuit diagram, or you can follow the steps given below.

• The 1.0kΩ resistor is banded brown–black–red–gold. The 100Ω resistor is brown–black–brown–gold.

- The first node is between the 9V (+) supply and the 1.0kΩ resistor (labeled as **R1**). Plug one lead (wire) of the 1.0kΩ resistor into any of the holes of the (+) bus at the upper edge of the breadboard. Leave the other lead unplugged for now. You have completed the first node.

- The second node is between the 1.0kΩ resistor (R1) and the long lead of the LED (labeled as **D1** for diode 1). Plug the resistor into a marked vertical 5-hole component strip. Plug the long lead of the LED into the same marked vertical component strip, while plugging the short lead of D1 into the unmarked component strip next to it. You have completed the second node. This is shown in figure 2-37.

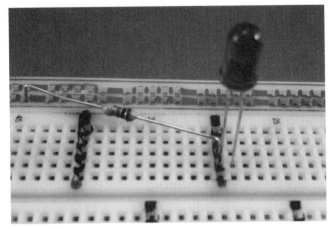

Figure 2-37.

- The third node is between the short lead of the LED (D1) and the first lead of the 100Ω resistor (labeled as **R2**). You already have the short lead of the LED plugged in. Plug one lead of the resistor into the same unmarked component strip. Leave the other lead of R2 unplugged for now. You have completed the third node.

- The fourth and last node is between the 100Ω resistor (R2) and the ground. Plug the second lead of R2 into any hole of the ground bus at the bottom. You have completed the final node.

Figure 2-38.

- When complete, the LED should light up. This circuit is shown in figure 2-38.

- Just as before, visualize the path the current is taking, and compare the circuit diagram to the real-world circuit on the breadboard. Trace the path of the current from the (+) terminal to the breadboard (+) voltage bus to R1 to D1 to R2 to the ground bus and then to the (−) terminal. This current flow is shown in figure 2-39. As you are doing this, notice that the current goes through each of the components one after the other.

- Clear your breadboard of components, ready to build another circuit.

This exercise shows that more than one resistor can be used in a particular signal path.

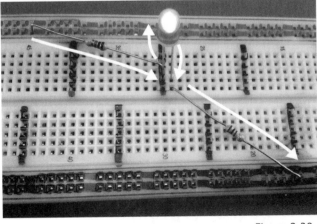

Figure 2-39.

Building the Third Circuit

This circuit uses an additional LED to demonstrate current flow.

- The diagram of the third circuit is shown in figure 2-40.

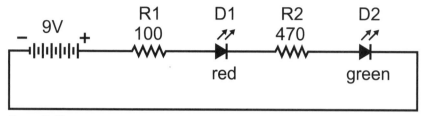

Figure 2-40.   Figure 2-40.

- Assemble this circuit on your breadboard. Follow the node-by-node procedure of assembling a circuit, just as you did for the first circuit. You can do this on your own by following the circuit diagram, or you can follow the steps given below. The 100Ω resistor is brown–black–brown–gold. The 470Ω resistor is banded yellow–violet–brown–gold.

- The first node is between the 9V (+) supply and the 100Ω resistor (labeled as **R1**). Plug one lead (wire) of the resistor into any of the holes of the (+) bus at the upper edge of the breadboard. Leave the other lead unplugged for now. You have completed the first node.

- The second node is between the 100Ω resistor (R1) and the long lead of the red LED (labeled as **D1** for diode 1). Plug the resistor into a marked vertical 5-hole component strip. Plug the long lead of the LED into the same marked vertical component strip, while plugging the short lead of D1 into the unmarked component strip next to it. You have completed the second node, as shown in figure 2-41.

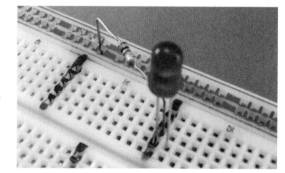

Figure 2-41.

- The third node is between the short lead of the red LED (D1) and the first lead of the 470Ω resistor (labeled as **R2**). You already have the short lead of the red LED plugged in. Plug one lead of the resistor into the same unmarked component strip. Leave the other lead of R2 unplugged for now. You have completed the third node.

- The fourth node is between the 470Ω resistor (R2) and the long lead of the green LED (labeled as **D2** for diode 2). Plug the resistor into a marked vertical 5-hole component strip. Plug the long lead of the green LED into the same marked vertical component strip, while plugging the short lead of D2 into the unmarked component strip next to it. You have completed the fourth node.

- The fifth node is between the green LED and the ground. Plug a piece of jumper wire with stripped ends into the unmarked component strip that the green LED is plugged into. Plug the other end of this wire into the ground bus at the bottom. You have completed the final node.

- When the circuit is complete, both of the LEDs should light up. This circuit is shown in figure 2-42.

Figure 2-42.

• Just as before, visualize the path the current is taking, and compare the circuit diagram to the real-world circuit on the breadboard. Trace the path of the current from the (+) terminal to the breadboard (+) voltage bus to R1 to D1 (red) to R2 to D2 (green) to the ground bus and then to the (−) terminal. This current flow is shown in figure 2-43. As you are doing this, notice that the current goes through each of the components one after the other.

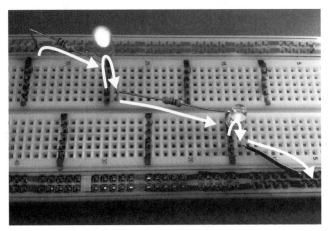

Figure 2-43.

• Clear your breadboard of components, ready to build another circuit.

This set of exercises has given you the basic skills in building a circuit on a solderless breadboard. You'll be using this skill throughout the rest of the book for building circuits to examine and experiment with.

3

The Digital Multimeter

Goal: When you have completed this chapter, you will be familiar with using the digital multimeter to measure voltage, current, and resistance.

Objectives

- Recognize the symbols for ohmmeter, ammeter, and voltmeter.
- Build circuits for demonstrating the use of the digital multimeter (DMM).
- Use the DMM to measure voltage sources.
- Use the DMM to measure voltage drops across components.
- Use the DMM to measure resistance.
- Use the DMM to measure current in a circuit.

Multimeter Background

In the early days of electronics, very large and expensive meters were needed to measure various electronic functions. An ohmmeter was used to measure resistance, a voltmeter was used to measure voltage, and an ammeter was used to measure current. This equipment generally used moving needle meters to display the output. With the advent of modern electronics, it became possible to integrate all of these meters into a single meter with a digital readout. Typically this unit is called a digital multimeter, abbreviated DMM, or a DVM for digital voltmeter. The DMM is the most common basic diagnostic tool used by electronics technicians.

Some meters are designed to perform a function called auto-ranging, which means that there is just one setting for each type of measurement. The meter figures out what the reading is and displays a prefix if needed. As cool as this may seem, I still prefer the DMM with manual ranges. It keeps me from having to keep track of whether there's a prefix or not, as the prefix on the display is usually pretty small, and it intrinsically gives me a sense of whether the voltage, current, or resistance is large or small.

Measuring Resistance

Resistance is the ability to restrict the flow of current. The DMM can measure resistance over a wide range. Its unit of measurement is the ohm, which is symbolized with the Greek capital letter omega (Ω). We'll start with measuring a 100Ω resistor. The bands of this resistor are brown–black–brown–gold. Later you will learn the resistor code, if you don't know it already.

Figure 3-1.

To measure resistance, turn the dial of the DMM to the resistance section, which is usually labeled with a Ω. Within this section choose the scale labeled **200**, as shown in figure 3-1. The reason you chose this setting is that it's the lowest scale that can measure the resistor. In general, you want the lowest scale possible, because it will give you the most precise answer. Certainly you could measure the 100Ω resistor on the 2M scale, which goes up to 2 million ohms, but 10,000Ω is the smallest resistance you can hope to see on that scale. A 100Ω resistor would appear to have no resistance, because it would give a reading of zero. It's like using the odometer of a car to measure the length of your dining room table: they both measure distance, but they are not even close to being on the same scale.

In addition, you'll need to check that the probes are inserted in the correct sockets of the DMM. The black probe is always inserted in the common socket, which is usually labeled **COM**. The red probe is a little more complex. Some models use the same socket for almost every function: DC voltage, AC voltage, resistance (shown by the Ω), and current (shown by A). Some models use different sockets for different functions, as shown in the figure. Determine which socket the red probe needs to go into for measuring resistance (Ω). Some meters have separate sockets for low current (mA) and high current (usually 10A).

Note that the meter reads **1** when there is no connection at all or when it is greater than the selected scale can display. In the case shown on the previous page, the meter can only read a resistance between 0 and 200. Any other resistance will give the **1** reading.

A meter for measuring resistance is called an ohmmeter, as we saw in the previous chapter. The symbol for the ohmmeter is shown in figure 3-2. To understand this symbol, we should imagine that there is a meter display located inside the symbol's circle, and that the two lines from the sides represent the two probe wires with the probes at their tips. Visualizing this will help us place the probes in the right places for taking measurements. This symbol does not simply indicate where to put a single probe. It shows the location of both of the probes and the meter itself.

Figure 3-2.

To measure the resistance of a component such as a resistor is easy: just put the red probe on one wire of the resistor and the black probe on the other wire. It doesn't matter which probe goes where, because resistance is the same either way. If you are measuring a resistor that is part of a circuit, there's an extra step: turn off or disconnect the voltage source before you start. Taking resistance readings on a resistor in a circuit when it is powered up will give you incorrect results, and it could possibly damage your DMM or the circuit.

When taking any measurement with your DMM, it's important to use enough pressure on the probes to make good contact between the probes and the wires. People have a tendency to use too light of a touch with the probe, perhaps not wanting to damage the component. You don't want so much pressure that you are in danger of bending the probe tips, but you want to be making sufficiently good contact to get a clear reading with the meter.

It's also a good idea to be in the habit of keeping your fingers away from the metal parts of the probes. With some readings, it's not going to make much difference if you are touching the metal or not. With other readings, being in contact with the probes could affect the reading, or could be harmful or even fatal to you. Rather than thinking about when it's safe and when it's not, it's better to just assume that what you are testing has the potential to kill you. Act accordingly by not making contact.

Measuring Voltage

There are two types of voltage the DMM can measure: DC voltage and AC voltage. Initially you'll be measuring the voltage from batteries, which are a source of DC voltage. You will learn how to set the meter appropriately to measure the voltage of a 9V battery. We'll discuss AC voltage and how it is different from DC voltage in a later chapter.

Turn the dial to measure DC volts. Different manufacturers use different labeling to indicate DC volts: the scale may be labeled **VDC, DCV,** or **V** with a solid line over a dashed line, as shown in the illustration. Since you're measuring a 9V battery, you want the smallest range that is large enough to cover it; on most DMMs, this is the 20V range, as shown in figure 3-3. Check that the probes are inserted in the correct sockets of the DMM for measuring DC voltage.

Figure 3-3.

The symbol for a voltmeter is shown again in figure 3-4. The voltmeter itself is shown by the circle and the meter's connections (the probes) by the lines at the sides. I keep mentioning that the probes are part of the symbol because students often ask where to put the other probe. People ask this because they think that the symbol indicates where to place one of the probes. The symbol doesn't just show where to put a single probe; it represents both of the probes and the meter itself.

Figure 3-4.

In the hands-on exercise for this chapter, you will be reading the voltages of several batteries and the voltage across a resistor in a working circuit. When you measure the voltage of a battery, start by placing the black probe on the negative (–) terminal of the battery. Follow this by placing the red probe on the positive (+) terminal. Watch the display, and hold the probes in position long enough for the reading to stabilize. How long this takes will depend on your DMM's design, but 10 seconds is usually long enough. If you're not getting a clean reading after 15 seconds, you are probably touching the probes too lightly on the terminals or wires. If that's not the problem, your DMM may be suspect.

Measuring Current

The symbol for the ammeter, for measuring current, is again shown in figure 3-5. Just as for the other meter symbols, it's important to regard the two lines at the sides as representing the two probes. Measuring the current in a circuit is just a little more complex than measuring voltage or resistance, because all of the current in the circuit must flow through the meter in order to be measured. In order to measure the current in a circuit, we start with a working (functional) circuit, open it at some point, and insert the meter into the circuit by placing the probes to complete the circuit again with the meter.

Figure 3-5.

This ammeter measures current by "counting" the electrons flowing through the meter. To do this effectively, the meter can't restrict the current flow at all, so it must act just like a piece of wire. There are many places in a circuit where we would not want to stick a piece of wire, so we must be more cautious with our ammeter placement than we are with our voltmeter.

In the hands-on exercise, you will connect a battery, a resistor, and an LED together, as shown in figure 3-6. This is the same circuit as the first one you built in the hands-on exercise in the previous chapter. The resistor and the LED provide a path for electrons to flow to the other battery terminal. This circuit diagram has an arrow indicating that the current is flowing downhill from the (+) terminal to the (–) terminal. Along the way, it's doing some work by using the LED to produce light.

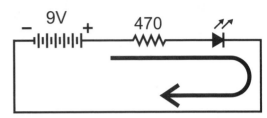

Figure 3-6.

A typical use of an ammeter is shown in figure 3-7. We have disconnected or opened the circuit, interrupting the current. We have then reconnected the circuit with the ammeter as part of the circuit. In this way, the current in the circuit is forced to flow through the ammeter. When the circuit is reconnected using the meter, the glow from the LED tells us that there is current flowing, because the LED is emitting light as a result of the flow of electrons. We can conclude that the electrons are also flowing through the meter, because there isn't any other way for them to get to the other terminal of the battery.

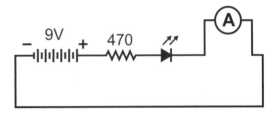

Figure 3-7.

This circuit diagram emphasizes the new path that the current takes through the ammeter by showing it going up into the meter and back down again. For ease of drawing, we usually show the meter as being just another component in the chain, as shown in figure 3-8.

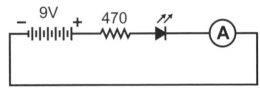

Figure 3-8.

Electrically the two schematics are identical: in each one the current flows from 9V battery (+) terminal through the resistor, through the LED, through the ammeter, and to the other battery terminal.

Using the DMM as an Ammeter

As well as measuring voltage and resistance, the DMM can measure current over a wide range. Current is how much electricity is flowing through a circuit. It is measured in amperes, usually shortened to amps. The amount of current in the circuits you will be studying is so small that it is measured in milliamps. A milliamp is 1/1000th of an amp, abbreviated **mA**.

To measure current, turn the dial to the DC current position, as shown in figure 3-9. Different manufacturers use different labeling: the scale may be labeled **ADC, DCA,** or just **A** with a solid line over a dashed line. Since you don't know how much current to expect, start with the highest range, which is usually 200mA, as shown. Some meters have a separate socket and range for measuring up to 10A. I've hardly ever used this range on my meter.

Figure 3-9.

Check that the probes are inserted in the correct sockets of the DMM for measuring DC current. Note that on some meters the red probe needs to be in a different socket.

It's essential to remember not to touch both of the battery terminals directly with the probes when the DMM is set for reading current. Think of the ammeter as a piece of wire with a digital display for measuring current. If you use a piece of wire to connect the terminals of a battery together, a great deal of current will flow, the battery will get very hot, and soon the battery will be drained. I don't recommend trying this to prove it to yourself, because it's a waste of a good battery. You may have encountered this accidentally if you have ever put a 9V battery in your pocket with loose change: a coin connects the terminals and the battery gets hot in your pocket. The meter's current-measuring circuitry is too delicate to survive high current flows, so it is protected by a fuse that will fail when too much current flows through it. When this happens, we say that the fuse is "blown." If your DMM does not give a reading or does not cause the expected functioning of the circuit when the DMM is connected, check that your circuit is correct by substituting a piece of plain wire for the meter. If your circuit works with the wire but not with the meter, check whether you need to replace the meter's fuse.

If the DMM is equipped with a Polyswitch self-resetting fuse or equivalent, you will simply need to wait a few moments for the fuse to reset. It's easy and inexpensive to modify the DMM for lab use by replacing the standard fuse with the resettable fuse, so that the fuse doesn't need to be replaced. This modification is described in Appendix C. When I'm teaching, I have all of my students change out their fuses to Polyswitches, so that their ammeters will still be functioning by the end of the class day.

The Accuracy of a DMM

When we take measurements of voltage or current or resistance, we always have to ask ourselves how accurate our measuring tools are. The DMM is really cool, because we put our probes in a particular spot and—"Presto!"—a little number flashes up on the screen to tell us what we want to know. The trouble is that a DMM costing $15 or $20 may not be particularly accurate; it gives us a number, but it may not be quite the right number. At the level of what we're working on, it's close enough. I wouldn't even mention this, except that you are about to add up two or three voltages and compare them to the reading you get from the voltage source. The slight inaccuracy of each of these readings means that the numbers are unlikely to add up perfectly. This is nothing to worry about. In general, a few tenths of a volt missing or extra in these calculations is normal. This is also why we don't get really excited about getting accurate calculations: if the calculator gives us an answer of 3.426623718V, we can comfortably round it to 3.4V.

One of the essential skills of using a DMM is to associate the property you are measuring with how to read it on the meter. For the first few hands-on exercises, I will say explicitly how to set your meter to make the measurements. Later on, I will simply ask you to measure the current, or measure the voltage, or measure the resistance. It will really help for to you to associate the property being measured with the setting on the meter and know how to connect the meter to the circuit to take the reading. This is the basic vocabulary of electronics.

Hands-On Practice

We'll start with measuring some batteries and resistors, and then we'll move on to taking measurements in a circuit. This exercise will give you the skills to use your digital multimeter (DMM) to measure the basic properties of electronic circuits: voltage, current, and resistance.

Materials

These are the materials you will need for these exercises.

- Digital multimeter (DMM)
- Breadboard
- 9V battery and battery clip with wires
- Several 1.5V batteries (AA or AAA cells, old or new)
- Four 100Ω resistors (banded brown–black–brown–gold)
- 470Ω resistor (banded yellow–violet–brown–gold)
- Four 1.0k Ω resistors (banded brown–black–red–gold)
- Two red LEDs
- Insulated solid wire (20, 22, or 24 gauge)
- Optional: grab bag of resistors
- Optional: dead piece of audio gear

Measuring Battery Voltages

- Set your meter to read DC voltage on the scale labeled **20V**, as shown in figure 3-10. Different meters may have somewhat varying formats, but they all work the same way. If the probes can be moved from socket to socket, check that the red probe is in the socket labeled **V.**

Figure 3-10.

- Read the voltage of the 9V battery. Place the black probe on the (−) terminal of the battery. Place the red probe on the (+) terminal. Hold the probes steady and firm on the terminals; you should be pressing hard enough to make good contact. Whenever possible, use the sides of the probes to make contact, not the tips. The surface area of the point of the tip is much smaller than the surface area of the sides, so establishing consistent contact with the metal surface of the terminal will be easier. Don't touch the metal parts of the probes with your fingers. The voltage present in a 9V battery is not going to do you any harm, but it's good to develop a habit of never touching the probes. Sometimes the voltage you are measuring is deadly, so it's better to think of it as always being deadly.

- Look at the reading displayed on the DMM. This is the voltage (electromotive potential) of the battery, measured in volts. The reading you get should be somewhere around 9.0V. If the reading doesn't stay stable after a few seconds, try pressing the probes harder against the terminals to keep the metal of the probe in full contact. The DMM uses a digital sampler to display the voltage. If the probe is scraping around on the surface, making contact and then breaking it before settling down in a new spot, the DMM won't be able to get a consistent reading.

- Next, read the voltages of your AA or AAA cells. Again, place the black probe on the (−) terminal and the red probe on the (+) terminal. You expect to see a reading of 1.4V or 1.5V on each battery. If you see a voltage reading of 1.0V or lower, the battery is no longer good and needs to be replaced. Remember to use the sides of the probes whenever possible and to use enough pressure to maintain a consistent reading. Now that you know how to check whether a battery is good, be sure to appropriately recycle any batteries that no longer have sufficient voltage.

• What you will find when measuring the voltage of batteries is that the actual voltage varies quite a bit from the listed voltage. When it's fresh, a 9V battery will usually have a voltage of about 10V. As this energy gets used up, the voltage of the battery drops to 9V, 8V, and so on. The voltage value printed on the battery is the called the *nominal voltage*.

This exercise has shown you how to measure the voltage of a battery with your DMM.

Measuring Resistance

The values of some 100Ω resistors will be the next measurements.

• Set your DMM for measuring resistance. Most DMMs have a setting labeled **200**. This is called the 200Ω scale, meaning that it will read any resistance value up to 200Ω. This is shown in figure 3-11. On most meters, the red probe goes in the same socket when measuring resistance and voltage. Check that the red probe is in the appropriate socket, labeled with the Ω symbol.

Figure 3-11.

• Measure the resistance of the resistor by setting the resistor on your work surface. Place the red probe on one lead of the resistor and the black probe on the other lead. Resistors are not directional, so the red probe can be on either end of the resistor. When you are taking these readings, you need to use enough pressure on the wires to get the probes to make good contact, just as you did with taking the voltage readings. For each of the resistors, you should get a clear reading of somewhere around **100.** For example, the resistor shown in figure 3-12 is giving a reading of **99.6Ω.**

Figure 3-12.

• Notice that there is some variation in the resistances of these components. Just as for the voltages of the batteries, 100Ω is the nominal resistance of these resistors. Their actual values can vary several ohms up or down from the nominal resistance value. The range of variation allowed is defined by the tolerance band, the last colored band on the resistor. We'll discuss this more when we work with the resistor color code.

- Now you will measure the resistance of the 1.0kΩ resistors. Set your DMM to the next scale up. On most meters this is the scale labeled **2k**. This is called the 2kΩ scale, as shown in figure 3-13. This scale can read up to 2kΩ (2000Ω). When you read the resistance this time, the meter will give you a reading somewhere around **1.00**.

Figure 3-13.

- When the meter is set on this setting or any setting with a "k" after it, it symbolizes adding a "kilo" prefix, meaning one thousand. If your meter reads **1.04**, your reading is 1.04kΩ. An example of such a reading is shown in figure 3-14, displaying a reading of **0.987kΩ**. Always remember to add a "k" to any reading the meter gives when it is set on this scale. When we remove the "k" from a number, we multiply it by 1000. In this way the **0.987kΩ** reading shown becomes 987Ω. Both of these are valid expressions for the same resistance. For more information about prefixes, consult Appendix A. It's a good idea to become very comfortable with adding and removing prefixes, especially the prefix for kilo. Many resistor values use this prefix.

Figure 3-14.

Rebuilding the First Circuit

You need a circuit for measuring voltage, current, and resistance. We'll start with the first circuit you built in the previous hands-on practice. You will need to set up your breadboard with a 9V battery, as before. This circuit is shown in figure 3-15. Assemble this circuit on your breadboard. If you need a reminder of how to do it, refer to the instructions in the previous chapter.

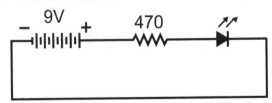

Figure 3-15.

Measuring Voltage in a Circuit

The goal of this exercise is to measure the voltage source and the voltage differences across the components.

- Set your voltmeter on the 20V scale, as you did earlier.

- Measure the voltage of the voltage supply by putting the black probe on the exposed wire of the ground bus and putting the red probe on the exposed wire of the (+) bus. This measurement is shown in figure 3-16.

- Record the result on a piece of paper. The reading should be around 9V.

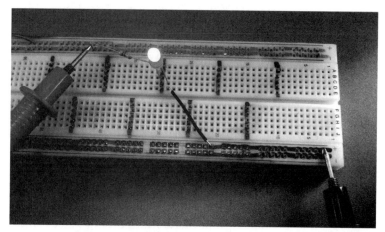

Figure 3-16.

- Measure the voltage across the resistor by putting the red probe on the side closer to the battery and the black probe on the other side. This measurement is shown in figure 3-17.

- Record the result on the piece of paper. The answer should be close to 7.5V.

- Because the resistor isn't a voltage source, we call the voltage the voltage drop across the resistor.

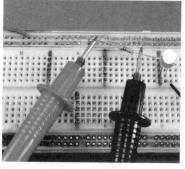

Figure 3-17.

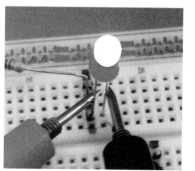

Figure 3-18.

- Measure the voltage across the LED, with the red probe on the lead (wire) closer to the (+) side of the battery and the black probe on the other lead (wire). This measurement is shown in figure 3-18.

- Record the result on the worksheet, and include the appropriate units! The answer should be somewhere around 1.5V.

This exercise shows that voltage is measured with the voltmeter across the two leads of the source or the component.

Measuring Current in a Circuit

The goal of this exercise is to measure the amount of current flowing in the circuit. Before you start, it is suggested that you modify your DMM with a self-resetting fuse, as shown in Appendix C.

- Just as you did when you built this circuit in an earlier hands-on exercise, visualize or imagine the current flowing through the circuit. It's starting at the battery and then running through the voltage supply bus, the resistor, the LED, the wire, the ground bus, and back to the battery. Tracing this route with your finger will help you imagine that there is current flowing through the circuit. It is this flow that you will be measuring with your meter.

- Set your meter to measure current, on the 200mA range. If your meter doesn't have the 200m range, try the highest current setting you can find (but not the super-high current 10A range). Some meters require that you plug the red lead into a different socket on the meter, the one labeled **mA**. Always leave the black lead plugged into the common (COM) socket. The meter is shown in figure 3-19.

- With the meter set for measuring current, it is vulnerable to excessive current flowing through it. Be especially careful with the probes of your meter when you have the selector set to current. Be very mindful of where you are putting your probes, of whether they are in a position where there is an unrestricted current supply across them. An example of what not to do is shown in figure 3-20.

Figure 3-19.

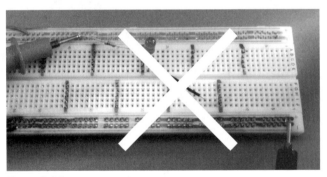

Figure 3-20.

- The circuit in figure 3-21 shows the ammeter as part of the circuit, so the current is forced to flow through the resistor, the LED, and the meter in turn. To get the ammeter to work, you must open the circuit and use the ammeter to complete it again.

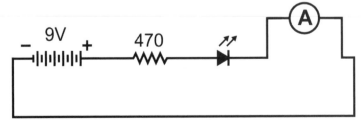

Figure 3-21.

- Unplug the end of the wire that is connected to the ground bus. Insert it into an unused column on the board. The LED should now be unlit.

- You will now use the ammeter to complete the circuit, in place of the wire you just removed. Just as the current flowed through the wire, it will now flow through the meter. Think of the meter as a piece of wire with a digital readout on it, because that's basically what it is in this situation. You must open the circuit and complete it again with the meter.

- Touch the red probe to the wire connected to the LED, pressing the side of the probe against the wire where it comes out of the breadboard. Touch the black probe to a wire coming out of the ground bus. The LED should light up again, showing that you have completed the circuit with the meter. Don't try to stick the probes into the holes on the breadboard; they don't fit and won't make contact. Instead, touch the side of the wire with the side of the probe. This is shown in figure 3-22. This is why I suggested that you insert the wire end into an unused column when you unplugged it. It's possible to make contact with the wire when it's waving around in the air, but it's a lot easier and you make better contact when it's plugged in.

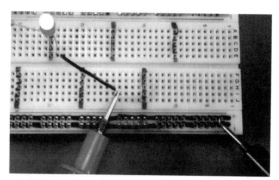

Figure 3-22.

- The current of the circuit is displayed. The current flowing in this circuit should be about 16 mA, so the display will read **016** or **16**. Since you are on the 200mA scale, the reading you get will be in mA, not amps (A).

- If no reading is displayed, check that your red probe is plugged into the correct socket. If there is still no reading and the LED doesn't light up, you may have blown the fuse, and you'll need to get another one or to replace it with a self-resetting breaker. You can confirm that the circuit is still good by reconnecting the piece of wire that you moved earlier to break the circuit.

- Write down your reading and its units on your piece of paper from earlier. Remember that you are on the mA scale. For the example given above, you would write down 16mA.

- Turn the selector dial of your DMM to the 20mA scale if it has one. On most meters, this is the position labeled **20m**.

- Repeat your current measurement as before: connect the red probe to the wire and the black probe to the ground.

- You should get a reading such as **15.9**. This indicates a current flow of 15.9mA. This is the same current you were measuring before, just a more accurate reading. The previous meter setting went up to 200mA, and this setting only goes up to 20mA. If we try to measure 30mA on this scale, the meter will display a **1** or the letters **OL** for overload. The advantage of this scale for us is better precision: we can detect the difference between 15.9mA and 16.4mA if we need to. On the 200mA scale, both of these measurements will display a reading of **16**.

- Clear your breadboard of components. Now you're ready to build another circuit.

This exercise shows that to measure current, the circuit must be opened and the meter inserted into the current path.

Measuring Resistance

The goal of this exercise is to measure the value of a resistor.

- Remove the 470Ω resistor from the breadboard.

- Set your DMM for measuring resistance by turning the selector to the **2k** setting in the Ω section. As you will remember from measuring resistor values in an earlier hands-on exercise, this scale will read any resistance up to 2kΩ (2000Ω).

- Check that the red probe of the meter is in the correct socket for measuring resistance, the one marked with Ω. This is shown in figure 3-23.

- Measure the resistance of the resistor by placing the red probe on one lead of the resistor and the black probe on the other lead. When you are taking these readings, remember that you need to use enough pressure on the wires to get the probes to make good contact. Resistance doesn't have directionality, so it doesn't matter where the red and black probes are, as long as one is on one lead and the other is on the other lead. This is shown in figure 3-24.

Figure 3-23.

- You should get a displayed reading of something close to **0.470**. Since you are using the 2k scale, you need to include a "k" prefix with the reading. You will also need the Ω unit for resistance.

- Write down the actual reading you get. For the display in figure 3-25, you would write down 0.464kΩ.

Figure 3-24.

Figure 3-25.

This exercise shows that for measuring resistance, the ohmmeter probes must be placed across the resistor, with no voltage applied to it.

Building the Second Circuit

We will use the second circuit to get some more practice measuring voltage, current, and resistance.

- Assemble the circuit shown in figure 3-26 on your breadboard. If you need a reminder of how to do it, refer back to the instructions of the previous exercise.

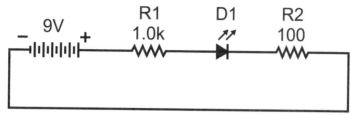

Figure 3-26.

Measuring Voltage in a Circuit

The goal of this exercise is to measure the voltage differences across the three components and become more comfortable with changing the meter to different measurement modes.

- Set your voltmeter on the 20V scale. Check that the red probe is plugged into the V socket. The black probe should be in the common socket as usual.

- Measure the voltage across the first resistor (R1) by putting the red probe on the side closer to the battery and the black probe on the other side. This measurement is shown in figure 3-27.

- Record the result on the piece of paper. The answer should be close to 6.1V.

- Measure the voltage across the LED (D1), with the red probe on the lead (wire) closer to the (+) side of the battery and the black probe on the other lead (wire). This measurement is shown in figure 3-28.

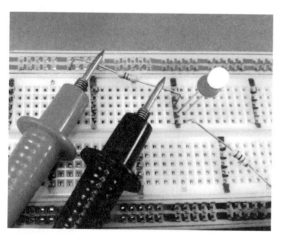

Figure 3-27.

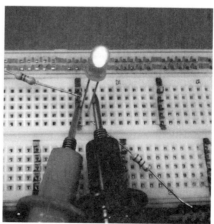

Figure 3-28.

- Record the result on the worksheet, and include the appropriate units! The answer should be somewhere around 1.8V.

Measure the voltage across the second resistor (R2) by putting the red probe on the side closer to the battery and the black probe on the other side. This measurement is shown in figure 3-29.

Figure 3-29.

- Record the result on the piece of paper. The answer should be close to 1.3V.

- On the piece of paper, add up the three voltages you just measured.

- Compare this total to the voltage of the supply, which you measured earlier for the first circuit. The total voltages of the three components should equal the voltage of the supply, or pretty close.

Measuring Resistance in the Circuit

The goal of this exercise is to measure the resistance of a resistor in a circuit. Up to now, all of our resistance measurements have been on resistors by themselves. The key point here is that we can't use an ohmmeter on a circuit that has current flowing through it. The reasons for this will be discussed in a future chapter. For the moment, all we need to know is that it won't work right.

- To measure resistance in a circuit, we need to turn off the current flow before we start. To do this, we can break the connection anywhere in the circuit. The easiest way is to break the last connection we made, which is the resistor that was plugged into the ground bus along the bottom of the breadboard.

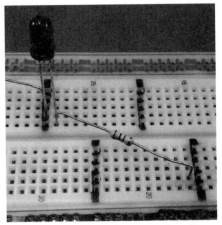

Figure 3-30.

- Unplug the resistor from the ground bus, and plug it into any unused component strip. This is shown in figure 3-30. The LED should no longer be lit.

- Set your meter as an ohmmeter for measuring resistance, on the 2k scale.

- Measure the resistance of the first resistor, R1, by putting the red probe on one lead of the resistor and the black probe on the other lead. Resistance is not directional, so the probes don't need to be in particular positions.

- Notice that this resistance measurement is made with no current running through the circuit.

- You should get a reading on the meter of something like **1.00**. Write down your actual reading with the appropriate prefix and unit.

- This reading is in kΩ, because you are in the Ω section of the meter on a scale labeled **2k**, so what you wrote down should be something close to 1.00kΩ.

- Measure the resistance of the second resistor, R2, in the same way: in parallel with no voltage across the resistor.

- You should get a reading on the meter of something like **0.10**. Write down your actual reading with the appropriate prefix and unit.

- This reading is in kΩ, because you are in the Ω section of the meter and still on the scale labeled **2k**. What you write down should be about 0.10kΩ.

- On your piece of paper, remove the "k" prefix and rewrite your reading without it. Your answer should be something close to 100Ω. For most purposes, this reading is fine. The rule of thumb is that you want to use the ohmmeter scale that gives you the most accuracy.

- Change your meter to the 200Ω scale for a more accurate measurement.

- Measure the resistance of the second resistor, R2, again.

- Write down your reading. This time you don't need to include a prefix, because you are on the on the 200Ω scale. You should get a reading between 95Ω and 105Ω.

This exercise shows that the resistance of components in a circuit can be measured with a DMM, but only when the voltage source is disconnected.

Measuring Current in a Circuit

The goal of this exercise is to measure the amount of current flowing in the circuit and to become more comfortable with changing the meter to different measurement modes.

- Visualize the current flowing through the circuit. It's starting at the battery and then running through the voltage supply bus, the first resistor, the LED, the second resistor, the ground bus, and back to the battery. Trace this route with your finger. It is this flow that you will be measuring with your meter.

- Set your meter to measure current on the 20mA range. Check that the red lead is in the socket labeled **mA**. Always leave the black lead plugged into the common socket. The meter is shown in figure 3-31.

- With the meter set for measuring current, it is vulnerable to excessive current flowing through it. Be especially careful with the probes of your meter when you have the selector set to current. I keep mentioning this because it is a common misstep for beginners measuring current. (I learned about this the hard way, by having to make several trips to the local electronics store for replacement fuses before I figured out why the fuse kept blowing.)

Figure 3-31.

- The circuit in figure 3-32 shows the ammeter as part of the circuit, with the current being forced to flow through the resistor, the LED, the second resistor, and the ammeter in turn.. To get the ammeter to work, you must open the circuit and use the ammeter to complete it again.

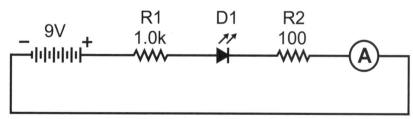

Figure 3-32.

- Unplug the resistor lead from the ground bus. Insert it into an unused column on the board. The LED should now be unlit.

- You will now use the ammeter to complete the circuit. Think of the meter as a piece of wire with a digital readout on it.

- Touch the red probe to the end of the resistor lead that you just moved, pressing the side of the probe against the wire where it comes out of the breadboard. Touch the black probe to a wire coming out of the ground bus. The LED should light up again, showing that you have completed the circuit with the meter.

- The current of the circuit is displayed. The current flowing in this circuit should be about 6.6 mA, so the display will read **6.60**. Since you are on the 20mA scale, the reading you get will be in mA, not amps (A).

- If no reading is displayed, check that your red probe is plugged into the correct socket. If there is still no reading and the LED doesn't light up, you may have blown the fuse, and you'll need to get another one or replace it with a self-resetting breaker. You can confirm that the circuit is still good by moving the resistor lead that you moved earlier to break the circuit, causing the LED to light up again.

- Write down your reading and its units on your piece of paper. Remember that you are on the mA scale. For the example given above, you would write down 6.60mA.

This exercise reiterates that current in a circuit can be measured by opening the circuit and inserting the meter.

Grab Bag Measurements and Dead Gear Measurements

If you have a grab bag or an assortment of resistors, I suggest that you take out a handful of them and practice measuring the resistance of each.

- The best place to start on your DMM is the middle range, usually 20kΩ.

- If you take a reading across a resistor and the meter display keeps showing **1**, turn up the meter to the next range and try again. If the meter shows a very small value, turn the meter to the next range down and measure again to get a more precise reading.

- This will give you some practice measuring a variety of resistors and getting used to using your DMM on different ranges.

- Another great place to use your DMM to measure resistance is inside a piece of gear. If you have a dead unit, open it up and look around. For these measurements, do not plug it in. Find the resistors on the circuit board (there are always some resistors), and practice measuring their values with your DMM.

You have now gotten some experience with using your DMM to measure voltage, current, and resistance, and you have built circuits on the breadboard. For later hands-on exercises, you may want to refer back to this exercise for the specifics of making these measurements. The instructions will not be as specific: for example, the directions might simply ask you to measure the current as shown in the circuit diagram, without giving you a setting or other directions.

4

Series and Parallel Connections, Ground

Goal: When you have completed this chapter, you will be able to recognize series and parallel connections and build circuits in these configurations.

Objectives

- Identify a series connection in a circuit diagram.
- Identify a parallel connection in a circuit diagram.
- Distinguish the difference between signal ground, earth ground, and chassis ground.
- Recognize the symbols for the three types of ground.

The Series Connection

A *series connection* is one in which a signal or current has only one path from one component to the next. An example of a series connection is how decorative light strings are often arranged, as shown in figure 4-1. If you have ever used strings of holiday lights for decoration, you have almost certainly encountered that frustrating set of lights that stops working when just one bulb burns out. The reason they behave this way is that they are wired in series: the current has to go through each little light in order to get to the end, and if any one single bulb is nonfunctional, the current stops and none of the lights work. This makes it very time consuming to find the burned-out bulb, because each bulb must be examined to find the bad one.

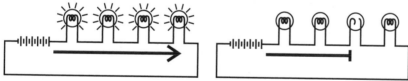

Figure 4-1.

All of the components in the circuit shown in figure 4-2 are connected in series. Another name for a series connection is an end-to-end connection. Each of the ends or terminals of the battery, resistors, LED, and ammeter are each linked to only one thing: the next component. In a series circuit there is only one path for the current to follow.

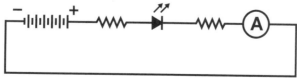

Figure 4-2.

The Parallel Connection

A *parallel connection* is one where the current has two or more choices of which way to go. As shown in figure 4-3, a typical application of a parallel circuit is any light circuit fixture in your house or office that has two light bulbs on the same switch. If either one of them burns out, the other one stays lit.

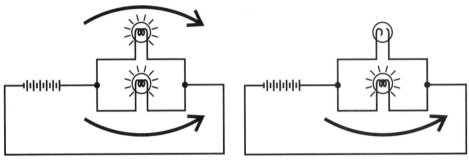

Figure 4-3.

Any circuit with more than one path is a parallel circuit. The easy way to recognize a parallel circuit is the *branch point*, which is marked by a dot at the junction. Figure 4-4 shows us some other examples of components in parallel circuits.

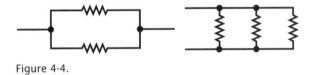

Figure 4-4.

The dots are important because they show that there's a connection between the wires. Always use a dot when two conductors connect, either in a T-shape or in a cross, as shown in figure 4-5. The T-shape is much preferred so that it can never be confused with two wires that cross each other but do not connect. The cross with a dot is still technically correct, but it is frowned on by most professionals. A cross with no dot indicates that there is no connection. Older circuit diagrams included a little hump when two wires crossed, to show that one jumped over the other. This little hump is no longer in use, but it may be found in circuit diagrams of old equipment.

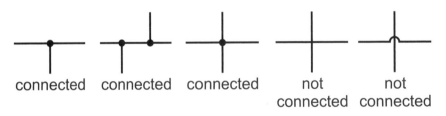

Figure 4-5.

The DMM in Series and Parallel

Now that we have the terms *parallel* and *series* to describe circuits, we can apply them to our DMM connections. Let's remember that the DMM has three basic metering functions: ohmmeter, voltmeter, and ammeter. Each of these measures a different property, and each has a different configuration.

- The voltmeter measures potential, or voltage. The meter is connected in parallel. We are measuring the potential between two points, as shown in figure 4-6.

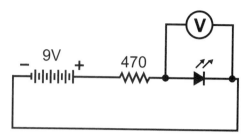

Figure 4-6.

- The ammeter measures current. The meter must be connected in series. We replace a wire in the circuit with an ammeter, which is a wire that also measures current. This is shown in figure 4-7.

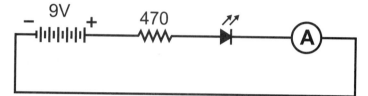

Figure 4-7.

- The ohmmeter measures resistance. The meter is connected in parallel. The circuit must not have current running through it, so that resistance between two points can be determined, as shown in figure 4-8.

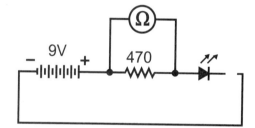

Figure 4-8.

The Three Types of Ground

A common term in electronics is the word *ground*. This can be a very confusing term, because it has three different meanings, and everyone has some preconceptions about its meaning. There are three types of ground in audio electronics: signal ground, chassis ground, and earth ground.

In audio electronics, *signal ground* is the most frequently used term. It's also the most difficult to understand, because it doesn't conform very well to what we usually think of as "ground."

Before we define signal ground and talk about how it is used, here is a little history. Early researchers into electricity observed that they could generate a strange force by rubbing certain things together, such as rubbing a piece of amber with a piece of wool. Further work found various ways of generating this "charge" and then storing it for a while in a device called a Leyden jar. This charge was thought to be a kind of invisible liquid. If it was given a metal path to the earth, it would flow down and be gone. The term *ground* became associated with "the place where electricity wants to go." Benjamin Franklin helped to further this thought by proving that lightning was electrical in nature. He developed the lightning rod to provide an easier path to the ground for the lightning strike.

Earth Ground

This brings us to the first of our ground definitions. An *earth ground* is electrically connected to the surface of the planet Earth, by way of a big piece of copper wire attached to a big copper rod pounded into the soil. The largest prong on a 3-prong plug is connected to earth ground. The function of the earth ground connection is to protect the equipment operator from electrical shock if a dangerous situation arises, by providing a preferred path for the current to the surface of the planet.

Figure 4-9.

Earth ground is important in audio electronics because it's what keeps the equipment safe, but it's not a direct part of the audio amplification circuit. The most common symbol for earth ground is shown in figure 4-9.

Chassis Ground

Another type of ground is called the *chassis ground*, usually represented by the symbol in figure 4-10. The chassis ground is a connection to the metal body or enclosure of the piece of gear. The chassis protects the circuit from RF (radio frequency) and other electrical "noise" by shielding it with a sheet of metal. It is almost always connected to earth ground, so that the user is protected if the chassis becomes electrically charged.

Figure 4-10.

Together the chassis ground and the earth ground are what we usually think of as the "ground" the current wants to flow to. In a sense, the chassis ground and the earth ground are just an extension of the surface of the planet.

Signal Ground

We regularly encounter equipment that works on electricity but doesn't have an earth ground. Just a few obvious examples are cellular phones, cars, jet airplanes, and laptop computers and tablets. In operation, none of these have an earth ground (though the image of what a Boeing 787 jetliner would look like trailing a ground cable from Tokyo to London is an amusing one). Clearly it isn't necessary for electronic equipment to have an earth ground in order to work.

Figure 4-11.

If we move a little closer to home, we just need to look at the circuit on the breadboard. It's powered by a battery, and we've seen that current is flowing in it and that it does work for us by the LED producing a little bit of light. Again, clearly no earth ground is involved here, because we never mentioned hooking it up to a wire that is connected to the planet's surface.

This leads us to our third type of ground, which is the *signal ground*. The most common circuit diagram symbol for signal ground is shown in figure 4-11. This ground is simply defined as being the place where the voltage is zero. This means that all other voltages are measured from this point, but it doesn't mean that all currents flow to this ground. To help us understand, let's go back a few chapters to the analogy of voltage being similar to the height of a water tank.

We looked at two water tanks, one 10 feet above sea level and the other 200 feet above sea level, and we said that the water in the higher tank had more potential. We just assumed that sea level was our starting point and that both of our tanks were being measured relative to that level. If you live by a part of the ocean near Seattle, for example, this would be a pretty easy assumption to make, because sea level is a constant in your life. (For simplicity, I'm ignoring the tides.)

Next, imagine taking a journey to Crater Lake in Oregon. This beautiful lake is at the top of a long-inactive volcano, at an elevation of about 6,200 feet above sea level. Everyone in the world uses sea level as the standard for measuring elevation. If we moved into the lodge at Crater Lake and stayed there for a year, we would probably stop thinking about elevations in terms of sea level. Instead we would be thinking about them in terms of the level of the lake surface. If we were to build a tank 200 feet above the surface of the lake and fill it with water, we would say that our tank had 200 feet worth of potential. It would be just like our 200-foot tank back down on the Seattle waterfront, but the whole system of lake and tank would be 6,200 feet higher.

We can also travel to the Dead Sea, which lies between Israel and Jordan. The level of this sea's surface is about 1,400 feet below regular sea level. We can camp on the shore and build a tank 200 feet above the level of this very salty body of water. When we pump water up into the tank, it now has 200 feet of potential relative to the surface of the Dead Sea.

How much work the water in the tanks can do is completely dependent on where we are standing and measuring from. Relative to the sea level, the water in the lake has 6,200 feet worth of potential. Relative to sea level, the tank 200 feet above Crater Lake has 6,400 feet worth of potential. If we go up to Crater Lake and think about the potential of the water in the 200-foot tank in Seattle, we realize that the water in that tank doesn't have potential: it is 6,000 feet below us. We're going to have to do a lot of work to get it up to our zero line at the lake surface. The water in the tank on the shore of the Dead Sea has 200 feet of potential relative to the surface of that body of water, but relative to the ocean, it has negative potential. We would have to pump it up 1,200 feet just to get it up to the zero line of sea level.

In the same way, we can define where our signal ground is for a particular piece of gear. As discussed earlier, signal ground is the reference point that we have established as being zero voltage, and we measure all voltages in the system from that point.

Voltage Supply Levels Relative to Signal Ground

Relative to the ground level, all voltages will be one of three types: positive (with current wanting to flow toward the ground), zero (with no current wanting to flow), or negative (with current wanting be pulled out of the ground toward that point).

Most electronics that run off a single battery use the negative terminal of the battery as the ground, so all voltages in the system are positive or zero, and all currents are flowing toward the ground. There is nothing magic about this arrangement: we can design and use a piece of electronics that does exactly the same job but defines the positive terminal as the ground, so all current in the system is being pulled out of the ground and down into the negative terminal. The work is exactly the same; just the names have changed.

Larger electronic systems such as power amplifiers and consoles use a *bipolar supply*, where there are identical positive and negative voltage supplies as well as the ground. For example, +18V and −18V is a typical voltage supply for a console. Power amplifiers usually need higher voltage supplies, such as +55V and −55V. For all of these, the signal ground is the zero point from which these voltages are measured.

We call it signal ground because it's also the zero point from which we measure the level of any audio signal we are working with. All audio electronics have connections consisting of two or more conductors. The most common audio connections are the TS (tip-sleeve, such as guitar cables), TRS (tip-ring-sleeve, such as balanced line cables) and XLR connectors (such as microphone cables). Each of these has one conductor dedicated to being the signal ground, the zero point for that signal. The changing voltage of our audio signal is present on the other conductor or conductors. The sleeve of the TS and TRS connectors is always the ground, and pin 1 of the XLR connector is always the ground.

In audio electronics, the term *ground* is understood to mean signal ground unless earth ground or chassis ground is specifically named as such. In the vast majority of gear, all the grounds are connected together. There are notable exceptions, and you should never just assume that the grounds are the same. If it's not connected to chassis or earth, the signal ground is said to be *floating*.

On some equipment, there is a switch that disconnects the signal ground from the other grounds. This is called the *ground lift* switch.

An unintended connection to the signal ground is called a *short circuit*. For an audio signal, this usually means that the signal is connected to the signal ground and thus is being held at zero volts instead of going where we want it to go. Later we will see that audio cable problems are initially diagnosed as either a short (meaning a short circuit) between two conductors or an open (meaning an open circuit), where there is no connection in a place where one is intended.

Using Ground Symbols in Circuit Diagrams

In the circuit diagram in figure 4-12, there are three resistors and a battery all connected in a loop. The connection between the 220Ω resistor and the (–) terminal of the battery is shown as a direct link.

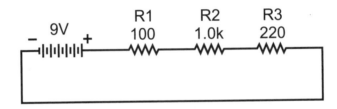

Figure 4-12.

Another way to depict this connection is to use the signal ground symbol, as shown in figure 4-13. Signal ground is the zero voltage reference point for the circuit, so by using this symbol, we are indicating that we will be using this as our starting point for measuring any voltages in the circuit. Even though there is no line drawn from the (–) side of the battery to the 220Ω resistor,

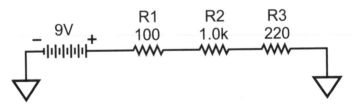

Figure 4-13.

there is still a connection between them, because they are both shown as being connected to the signal ground. This is a common way of showing a circuit. Large circuits can have dozens or hundreds of components that have a connection to signal ground. It's a lot easier to show these connections with signal ground symbols than to draw lines everywhere. Physically, there is a piece of conductive metal that connects them all, but we don't go to the trouble of drawing it out. We use the shorthand of the signal ground symbol instead.

Figure 4-14 also uses the signal ground, but the shorthand of the signal ground symbol is taken one step further. The (–) terminal of the battery is not shown at all. In this circuit diagram, the (+) terminal of the battery is shown with its voltage level, and it is taken for granted that the (–) terminal is connected to signal ground. This is common practice in audio gear, and it is the convention that will be followed for the rest of the book.

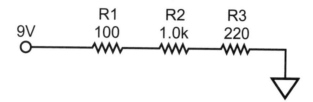

Figure 4-14.

Unfortunately the usage of the three symbols for ground is by no means standardized, so schematics can vary. I have seen each of these symbols used for signal, earth, and chassis grounds. In most cases it doesn't matter, because all of the grounds are connected together. In the few situations I've seen where there were distinct signal, earth, and chassis grounds, the schematic made the situation clear.

The DMM and Signal Ground

Now that we have a clear definition of signal ground, we will be using it frequently when talking about our measurement technique. When measuring voltage, we will always be placing the black probe of the DMM either on the ground or in the direction of the ground. The circuit diagram

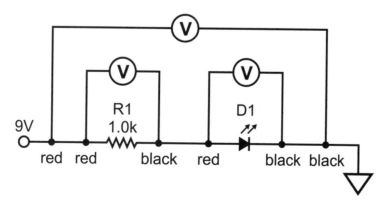

Figure 4-15.

doesn't show this. It's just implied by the use of a voltmeter. Figure 4-15 shows three voltage measurements in a circuit: the total voltage, the voltage across resistor R1, and the voltage across the LED D1. In each case the black probe is either on the ground or closer to the ground. In the future we won't see the black probe labeled explicitly like this, because it's just assumed that the

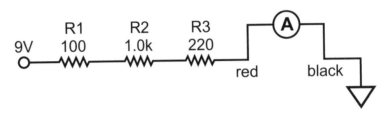

Figure 4-16.

black probe is on the ground or as close to it as it can get.

There is a similar situation for measuring current. In all of the current measurements you have made so far, you have disconnected the circuit at the ground bus and used the meter to reconnect the circuit to the ground. The current has then flowed through the circuit, into the red probe,

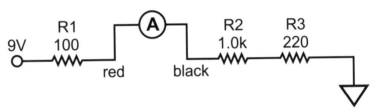

Figure 4-17.

through the meter, out the black probe, and to ground. This is shown in figure 4-16.

We're not limited to making current measurements at this point. We can open the circuit anywhere and insert an ammeter to read the current. An example of this is shown in figure 4-17. Again, in the future we won't see the red and black probes labeled explicitly, because it's just assumed that the black probe is either on the ground or as close as possible to the ground. When the meter is positioned this way, the ammeter should give a positive reading. If the reading is negative, the probes are probably reversed with the red probe closer to the ground.

Hands-On Practice

This exercise will explore building the parallel circuit and contrasting its behavior with the series circuit. This exercise will also continue to explore your use of the digital multimeter (DMM) to measure voltage, current, and resistance in different places in the circuit.

Materials

These are the materials you will need for these exercises.

- Digital multimeter (DMM)
- Breadboard

- 9V battery and battery clip with wires
- 100Ω resistor (banded brown–black–brown–gold)
- 470Ω resistor (banded yellow–violet–brown–gold)
- 1.0kΩ resistor (banded brown–black–red–gold)
- 4.7kΩ resistor (banded yellow–violet–red–gold)
- 22kΩ resistor (banded red–red–orange–gold)
- Red LEDs
- Green LEDs
- Insulated solid wire (20, 22, or 24 gauge)
- 2 jumper cables with alligator clips

Building the Parallel Circuit

Up to this point, all of the circuits you have built have been series circuits, with just one path for the current. This is the first parallel circuit in the hands-on exercises, so there are a couple of tricks to watch for.

- Set up your breadboard with a 9V battery, as before.

- The circuit is shown in figure 4-18. To assemble this circuit, we'll take the same node-by-node approach that we used in earlier circuit construction. Remember that a node is simply the name we give to a connection between two or more components. We could choose to start anywhere in the circuit as long as we completed all of the nodes, but it's easiest here to start with the 9V supply.

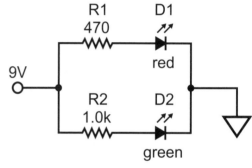

Figure 4-18.

- In the circuit diagram, we see that there are two components connected to the voltage supply: R1 (the 470Ω resistor) and R2 (the 1.0kΩ resistor).

- Find these two resistors based on their color banding. R1 is banded yellow–violet–brown–gold, and R2 is banded brown–black–red–gold.

- Plug both of these resistors into the (+) voltage bus along the top of the breadboard. Leave the other ends unplugged and free for now. You have just completed the first node.

- It's not obvious which is the second node, so we're just going to choose one: the free end of R1. This second node connects to only one other component: the (+) lead of the red LED, called D1.

- Plug the free end of resistor R1 into a marked vertical 5-hole component strip. Plug the long lead of red LED D1 into the same marked vertical component strip, while plugging the short lead of D1 into the unmarked component strip next to it. You have completed the second node.

- We'll choose the free end of R2 for the next node. This third node connects to one other component: the (+) lead of the green LED called D2.

- Plug the free end of resistor R2 into a marked vertical 5-hole component strip. Plug the long lead of green LED D2 into the same marked vertical component strip, while plugging the

Figure 4-19.

short lead of D2 into the unmarked component strip next to it. You have completed the third node. This is shown in figure 4-19.

- The fourth node is the connection of the (−) leads of both D1 and D2 to the ground.

- Use two pieces of wire to connect the unmarked component strips containing the short leads of the LEDs to a marked component strip. Use a third piece of wire to connect this component strip to the ground bus at the bottom of the board. Both of the LEDs should light up. This is shown in figure 4-20. (You could have plugged the two wires directly into the ground, but you are building the circuit this way to make it easier to measure the current later.)

- A parallel circuit is one where there is more than one path for the signal to follow. You can prove to yourself that this is a parallel circuit by unplugging resistor R1 (the 470Ω resistor) from the supply. The red LED of the unplugged arm will go out. The green LED in the other arm of the circuit will stay lit. This demonstrates that the current in the two arms is independent.

- Plug the resistor back in so that both LEDs are lit again.

- Just as you did when you built circuits earlier, visualize the current flowing through the circuit. This is a little more complex than what you've imagined before, because the current splits and comes back together. Think of it as starting at the battery and then running through the voltage supply bus, splitting to flow through the two resistors and then through the two LEDs, through the two wires to the connection strip, through the single wire, through the ground bus, and back

Figure 4-20.

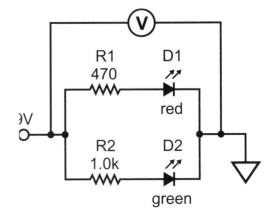

Figure 4-21.

to the battery. Try tracing this route with your fingers.

Measuring Voltage in the Parallel Circuit

This part of the exercise revisits the technique of measuring the voltage drop across a component, and it demonstrates that the voltage drops add up to the voltage of the source.

- Set your DMM for reading DC voltage on the 20V scale.

- Measure the voltage of the voltage supply. Notice that here we assume that the point we are measuring from is the signal ground, as shown in figure 4-21. This means that the black probe goes on the ground bus, and the red probe goes on the voltage supply bus. You should get a reading of around 9V. (You have made this measurement several times before, and you will certainly need to make it again. It's a good routine to confirm what the supply voltage is before embarking on other voltage measurements.)

- Write down this voltage.

- Measure the voltage drop across R1. Unlike the previous measurement, where we measured from the ground to another place, this time we are measuring the voltage on one side of the component compared to the other. The black probe goes on the side of the resistor closer to the ground. This measurement is shown in figure 4-22. You should get a reading around 7.2V.

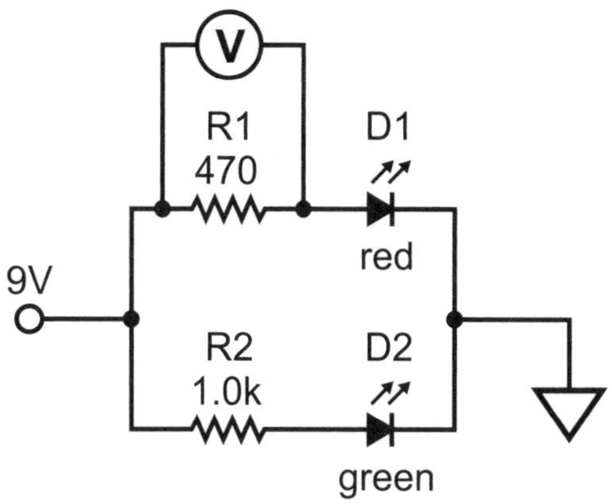

- Write down this voltage.

- Measure the voltage drop across D1, the red LED. Just as with the resistor, measure with the black probe closer to the ground (in this case it's on the ground.) You should get a reading around 1.8V.

- Write down this voltage.

- Add up the two voltages that you just measured, across R1 and D1. The sum of these two voltage drops should total the supply voltage that you measured earlier.

- Measure the voltage drop across R2, the 1.0kΩ resistor. You should get a reading of about 7.0V.

- Write down this voltage.

- Measure the voltage drop across the green LED, D2. You should get a reading around 2.0V. Write down this voltage.

- Add up the two voltages that you just measured, across R2 and D2. The sum of these two voltage drops should total the supply voltage that you measured earlier.

- Notice that even though the supply voltage is the same across these two arms of the circuit, the voltage drops across the two resistors are somewhat different. This is because the voltage drops across the LEDs are different. LEDs of different colors have some variation in their voltage drops. We'll see more about this in the chapter on semiconductors.

Figure 4-22.

Measuring Current in the Parallel Circuit

This exercise will show you that separate currents flow in the arms of a parallel circuit. It also demonstrates how the meter displays an overload when a current is too large.

- Set your DMM for reading DC current on the 200mA scale.

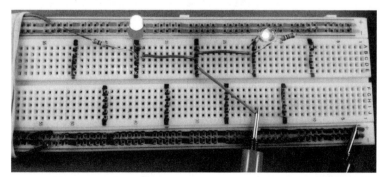

Figure 4-23.

- Unplug the piece of wire connecting the circuit to the ground bus, and insert the end in an unused component strip.

- Use the DMM to measure the current with the red probe on this wire end and the black probe on the ground. This is shown in figure 4-23.

- The equivalent schematic is shown in figure 4-24.

- You should get a reading of about 22.1mA.

- Write down this reading.

Figure 4-24.

- Use one of the jumper cables with alligator clips to connect the black probe to the ground, by attaching one alligator clip to the metal of the black probe and the other alligator clip to the wire loop of the ground point on the breadboard..

- Use another jumper cable to connect the red probe to the open wire end where you were measuring earlier. These two jumper-cable connections are shown in figure 4-25. When this connection is made, the LEDs should turn on.

- Look at your ammeter reading. It should be the same reading that you got earlier.

- You now have the ammeter set up so that you can read current without having to move the probes, because they are already connected into the circuit.

- Unplug the wire that connects the red LED D1 to the connection strip. The red LED will go

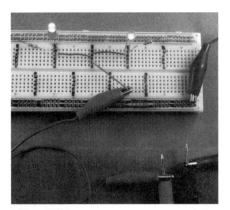

Figure 4-25.

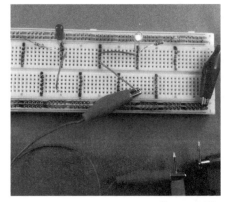

Figure 4-26.

out, and the current will decrease to about 7mA. The green LED will remain lit. What you are measuring is the current through the "green" arm of the circuit. This measurement is shown in figure 4-26.

- Write down this new current reading.

- Plug the wire back in so that both LEDs are lit.

- Unplug the wire that connects the green LED D2 to the connection strip. The red LED will remain lit, and the green LED will go out. The current will decrease to about 15mA.

- Write down this new current reading.
- Plug the wire back in so that both LEDs are lit.
- Add together the two most recent current readings, for the first arm and the second arm. These should add up to the first current reading you took.
- Leaving the meter in place, remove the 470Ω resistor R1.
- In its place, insert a 22kΩ resistor (R3). The schematic is shown in figure 4-27. Connect it exactly as R1 was connected, with one lead of the resistor in the voltage supply bus and one lead connected to the long lead of the red LED, in the marked connection strip. You do not expect the red LED to light, but the green LED should still be lit.

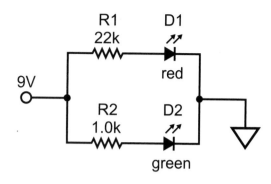

Figure 4-27.

- Your ammeter should be reading about 7mA.
- The measurements that follow will involve smaller currents, so change your DMM to the 20mA range.
- Write down the reading from this new setting. Do not be concerned if it is a little higher or lower than what you were reading before.
- Unplug the wire that connects the green LED D2 to the connection strip. The green LED will go out. The red LED is still unlit. The current will decrease to about 0.3mA.
- Write down this new current reading.
- The interesting observation here is that the meter is showing us that there is current flowing through the resistor R3 and the red LED D1, but there is no light from the LED. We can conclude that the meter can show us current flow even when we can't see it in other ways. We can also conclude that LEDs are not always good indicators of current: if there is not enough current, the LED will not light.
- Change your ammeter to the 2mA scale.
- You should be still seeing a reading of 0.3mA.
- Plug the wire back in so that the green LED D2 is lit.
- By plugging the wire back in, the current is now back up to about 7mA. On its present setting, the meter has a maximum readout of 2mA, so it can't display the 7mA. Most meters show a **1** or **OL** to indicate an overload. Whatever your meter is displaying, you should remember it, because it means that you need to turn the meter to a higher scale.
- Turn the meter back to the 20mA scale.
- Unplug the wire connected to the red LED, even though it isn't lit.
- When you do this, you should see the current drop a little bit.
- Write down the present reading. This is current through the green LED.
- Add together the current you just observed with the current you read for the red LED.

- This total should be the same as the reading you got when you first changed the resistor to a 22kΩ and changed the meter setting.
- Disconnect the meter. Plug the wire back in to restore the circuit to its former condition. Remember that the red LED will not be lit, but the circuit must still be complete, as shown in figure 4-28.

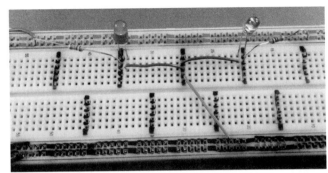

Figure 4-28.

Measuring Resistance in the Parallel Circuit

This exercise reminds us to disconnect the voltage source before taking resistance measurements, and shows how the meter displays an overload when the resistance being measured is too large.

- Set your DMM for reading resistance on the 2kΩ scale.
- Unplug and remove the piece of wire connecting the circuit to the ground bus. Both LEDs will be unlit.
- Use the DMM to measure the resistance of R2, the 1.0kΩ resistor. You should get a reading of about **1.00** on your meter, indicating 1.0kΩ.
- Use the DMM to measure the resistance of R1, the 22kΩ resistor. You should get a reading of **1** or **OL**, indicating that the value is too big to read. This is just like the **1** or **OL** in the current reading, which indicated that the scale was too small.
- Take a moment to compare these two readings of **1**. The first shows that the resistance is 1.00kΩ, and the other shows that the meter needs to be adjusted down to take a valid reading.
- Turn the meter up to the next scale, which is 20kΩ. The name of the scale indicates that it can read up to 20kΩ.
- Try taking a reading—you will see the **1** or **OL** on the display. This reading is indicating that you need to turn the meter up to the next scale.
- We are looking at measuring the 22kΩ resistor R3, and the scale doesn't go that high, so you need the next scale.
- Turn the meter scale up to the 200kΩ scale.
- You should see a reading of **22.0,** indicating resistance of 22kΩ.
- Clear the breadboard for building the next circuit.

Measuring Voltage Relative to the Ground in a Series Circuit

So far, all of the voltage measurements you have made have been across either the voltage source (the battery) or the component (the resistor or the LED). In this exercise, you will learn how to measure voltage at different points of a circuit relative to the ground.

- This circuit is shown in figure 4-29.

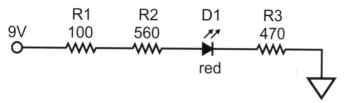

Figure 4-29.

- In the circuit diagram, you can see that you need three resistors: R1 (the 100Ω resistor), R2 (the 560Ω resistor), and R3 (the 470Ω resistor). Find these three resistors based on their color banding. R1 is banded brown–black–brown–gold, R2 is banded green–blue–brown–gold, and R3 is banded yellow–violet–brown–gold.

- Use these resistors and the red LED to build the circuit. There aren't specific instructions here, because this is a simple series circuit, and you have already built several of these. All you need to do is identify each node in turn and identify what two components are connected together. Remember that the marked component strips are good places to make connections between two components. Any component strip will work as long as the leads of both components are plugged in, but the marked ones are easier to use.

- When you are done with your circuit, the LED should light up.

Figure 4-30.

- Visualize the current flowing through the circuit: from the battery, to the voltage supply bus, through R1 (the 100Ω), through R2 (the 560Ω), through the LED, through R3 (the 470Ω), to the ground bus, to the battery. Trace this path with your finger. If the current is not going in this order, check the schematic and figure out which components are in the wrong places. The correct circuit is shown in figure 4-30.

- Set your DMM for reading voltage on the 20V scale.

- With your black probe on the ground, use the red probe to measure the voltage of the voltage supply. This is shown in figure 4-31. The reading should be about 9V.

- Write down the actual voltage, labeled as **position 1**.

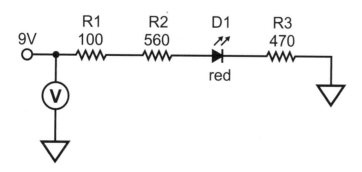

Figure 4-31.

• Leaving the black probe on the ground, use the red probe to measure the voltage at the next node, down on the breadboard between R1 and R2. Check that the probe is making contact with one of the leads. This is shown in figure 4-32. The reading should be about 8.3V.

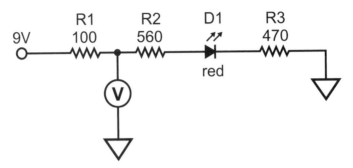

Figure 4-32.

• Write down the actual voltage, labeled as **position 2**.

• Leave the black probe on the ground. Move the red probe all around the leads of the node you just measured. You will notice that the voltage is the same (~8.3V) at every location, from where the lead exits the body of the 100Ω resistor (R1) down to the breadboard, up the other lead from the breadboard to where the lead enters the body of the 560Ω resistor (R2). This shows us that all of these points are electrically equivalent, meaning that they are all at the same voltage level relative to the ground. We can use any of these points as our measurement point for this node.

• Still leaving the black probe on the ground, use the red probe to measure the voltage at the third node, between R2 and the LED D1. As above, you should find that putting the probe at any point on the lead of R2 or the long lead of D1 gives you the same reading. This voltage should be about 4.7V.

• Write down the actual voltage, labeled as **position 3**.

• Again still leaving the black probe on the ground, use the red probe to measure the voltage at the next node, between the LED and R3. It should be about 3.0V.

• Write down the actual voltage, labeled as **position 4**.

• Still leaving the black probe on the ground, measure the voltage at the last node where the lead from last resistor R3 enters the ground bus. As we should expect, this reading will be **0V**. Taking this reading is a somewhat silly thing to do, as we don't expect to see any voltage, but it makes the point for us that we know that this lead of the resistor really is part of the ground, because it's connected directly to the ground bus.

• Electrically, the ground extends from the (–) terminal of the battery to all of the wire on the ground side of R3. We won't see any voltage at any point on the ground, and we could use any of these points as the place to put our black probe so that we can measure voltage at other places in the circuit.

• This reminds us what we said about signal ground in the chapter text: we defined it as the point we measure all voltages from, which we have defined as being zero volts. It is our sea level for this circuit.

This part of the exercise has demonstrated how to measure voltage levels at different points in the circuit, all relative to the ground. Next we will revisit the voltage drop.

Measuring Voltage Drops in a Series Circuit

The earliest voltage measurements in these exercises were the voltages across components. We called these *voltage drops* because each is a voltage difference across a component that is not a voltage source such as a battery. In this circuit, you can measure voltage across each of the components.

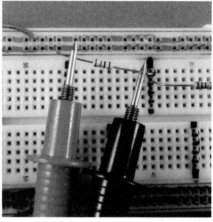

* Leave your DMM set for measuring voltage on the 20V scale.

* With the black probe of the meter closer to ground, take a voltage measurement directly across resistor R1. You expect a voltage of about 0.6V. This is shown in figure 4-33.

* Write down the actual voltage, labeling it R1.

* Take the voltage measurement directly across R2. You should see a voltage of about 3.6V.

Figure 4-33.

* Write down the actual voltage, labeling it R2.

* Take the voltage measurement directly across the LED D1. You should see a voltage of about 1.7V.

* Write down the actual voltage, labeling it D1.

* Take the voltage measurement directly across R3. You should see a voltage of about 3.1V.

* Write down the actual voltage, labeling it R3.

* You have now taken voltage measurements of all of the individual components while they are operating in a series circuit. In the next chapter, we will see how this part of the exercise relates to Kirchhoff's voltage law.

Adding a Resistor in Parallel to an Existing Circuit

We have one last objective, which is to add another resistor to the circuit. This resistor will be in parallel with one of the other resistors.

* This circuit is shown in figure 4-34.

* Your circuit should still be working, with the LED lit.

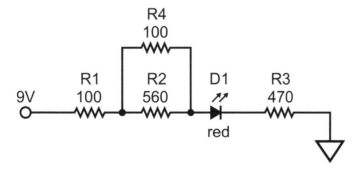

Figure 4-34.

- You will be adding an additional 100Ω resistor, R4. Find this resistor based on its color banding: brown–black–brown–gold.

- You are adding this resistor in parallel to R2.

- Plug one end of the new resistor R4 into the component strip of the second node, where R1 and R2 come together.

- Plug the other end into the third node, where R2 and D1 are connected.

- When you make this connection, you should see the light of D1 get slightly brighter. This shows that current is flowing through both of the resistors. The circuit is shown in figure 4-35.

Figure 4-35.

- Optional: Set up your DMM to measure current, and prove to yourself that the current really does increase when the resistor is added.

These exercises have given you experience with building series and parallel circuits, and with measuring their properties.

5

Kirchhoff's Laws, Current Convention, and Ohm's Law

Goal: When you have completed this chapter, you will have a working knowledge of Ohm's law and Kirchhoff's laws, which show us the relationships between voltage, current, and resistance.

Objectives

- Understanding the voltage drop.
- Understanding Kirchhoff's voltage and current laws.
- Understanding the difference between conventional current and real-world current.
- Understanding Ohm's law.

The Voltage Drop

Every circuit has a voltage source, and every circuit has one or more components connecting the two terminals of the voltage source and providing a path for the current. For each of these components, there is a higher voltage on one side than on the other, so that current flows "downhill" from high to low. This voltage difference across a component is called a *voltage drop*.

In the hands-on exercises, you have been measuring the voltage drops across resistors and LEDs. Any measurement where the black probe is on one lead of the component and the red lead is on the other is a measurement of a voltage drop. Toward the end of the most recent exercise, you measured four voltage drops in a row: resistor R1, resistor R2, LED, and resistor R3.

Kirchhoff's Voltage Law

In the mid-1800s, a German physicist named Gustav Robert Kirchhoff did a series of experiments with electrical circuits. From these experiments, he developed two very sensible conclusions about the properties of electricity—specifically, the properties of voltage and current, which are now known as Kirchhoff's Laws. The essence of these laws is that the universe is a balanced place, and voltage and current also want to be in balance. First we will examine Kirchhoff's voltage law: "The algebraic sum of the electromotive forces and the voltage drops around any closed circuit is zero." While it sounds pretty scary when stated in this formal way, the meaning of this law is quite simple. Kirchhoff is saying that voltage can't just magically appear in a circuit, nor can it magically disappear. We can also shorten this law to "Voltage sources = voltage drops." The voltage source or sources must be balanced by equivalent voltage drops.

We already saw an example of how this works when we took voltage readings of the voltage source and the components in the hands-on exercise. Most recently, there was a circuit with three resistors (R1, R2, and R3) and an LED (D1). The electromotive force (the source—i.e., the battery) was measured at 9V. The voltage drop across R1 was 0.6V. The voltage drop across R2 was 3.6V. The voltage drop across D1 was 1.7V. The voltage drop across R3 was 3.1V. The voltage drops as we go around the circuit add up to a total of 9V, but since they are voltage drops we say that their total is -9V. Together the battery and the voltage drops add up to zero: 9V + (-9V) = 0V. Earlier we expressed the voltage drops as positive numbers, and this is typical practice. If we wanted to be strictly correct, we would express them as -0.6V for R1, -3.6V for R2, -1.8V for D1, and -3.0V for R3.

If all of these numbers are making your head spin, relax. All we need to know is that voltage drops add up to voltage sources, and that's all based on adding up voltages. If we added up the voltage drops and found that they didn't equal the voltage source, we would need to look through the circuit and find the "missing" voltage. It has to be there somewhere, because Kirchhoff's voltage law says that it can't just disappear. Similarly, potential can't just arise from nothing.

Wes used this simple example to illustrate Kirchhoff's voltage law, but it can be applied equally to the most complex circuit. A consequence of Kirchhoff's voltage law is that if we include more components in the circuit, the voltage drops will be spread around between them. If we remove components, the voltages will also redistribute themselves. For example, we could remove R3, the 470Ω resistor from the circuit, as shown in figure 5-1. The voltage drop across R1 would jump to 1.1V. Across R2 we would now see 6.1V, and we would still see 1.8V across the LED D1. 1.1V + 6.1V + 1.8V = 9.0V, which is what we expect.

Just as before, these voltages add up to exactly 9V. There are equations that let us calculate what these individual voltages are, but we're not concerned about them at the moment. The important point is that however many components we have in the circuit, the

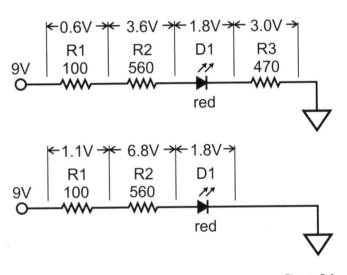

Figure 5-1.

voltages get distributed among them. (You may have noticed that the LED always stays the same, at 1.8V. This is correct. There is a reason for it, but we're not going to address it until we get to the first chapter about semiconductors.)

Conventional Current

Figure 5-2 shows the simple circuit of a 9V supply, a 470Ω resistor, an LED, and an ammeter. You'll notice that the diagram shows the flow going from (+) to the ground, which is connected to the (−) of the battery. Earlier when we were looking at the structure of the atom, we saw that electrons are negative. We said that the negative electrons are trying to get back to their home atoms, so we would expect them to flow from (−) to (+). In the real world, this is exactly what is happening.

The phenomena of the flow of electricity were observed long before electrons were discovered, and it appeared that positively charged particles were moving from (+) to (−), so this model became the accepted one. Benjamin Franklin is credited with being the first to assign the positive and negative labels to the forces of electricity.

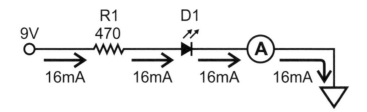

Figure 5-2.

The positive-to-negative current is called the *conventional current*. More than 125 years passed before the work of physicist J. J. Thomson showed that the negatively charged electron is the particle that is moving. By then it was hard to get the books rewritten, and the tradition of conventional current continues to this day.

In any case, the diagram in the earlier figure shows the conventional view of electricity. It's not really the way that electricity works, but nobody except transistor designers and physics purists worry about it very much. In this book, and in most electrical engineering classes, electricity is thought of as a flow of positive particles that moves from (+) to (–). Occasionally in this book we'll talk about the true flow of electrons, but most of the time we will follow the conventions established over 260 years ago.

Kirchhoff's Current Law

Kirchhoff has a scary-sounding law for current also: "For any point in a closed circuit, the algebraic sum of the current entering that point and the current leaving that point is zero." While stated in a formal fashion, this law is also pretty easy. The current going into a particular point has to come back out, and this is true for every point in the circuit. We can simplify this law also: For any point, current in equals current out. Again we see the theme of balance: the current entering a point balances the current leaving the point.

With the simple circuit shown above, we can use this law to show that the current is the same at every point in the circuit. Later we'll be working with circuits that have branching and we'll have more opportunities to use Kirchhoff's current law.

The resistor acts to restrict the amount of current flowing through the circuit. Without the resistor in place, the current through the LED would be very high, which would destroy the LED. It is tempting but incorrect to think of there being a large flow on the left side of the resistor, and the resistor pinching it down and only letting through a trickle on the right side. If that were true, it would violate Kirchhoff's current law, because the resistor would have more current entering it than leaving it. The consequence is that we can put the resistor anywhere in the circuit, and it will do the same job of restricting the current to a certain level. This is shown in figure 5-3.

You saw an example of Kirchhoff's current law in the parallel circuit you built for the hands-on exercise. You measured the total current flowing in the circuit at about 22mA. Further measurement showed that the current flowing in one arm of the circuit is about 15mA, and the current flowing in the other arm is about 7mA. Kirchhoff's current law tells us that we should expect that these two

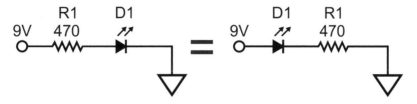

Figure 5-3.

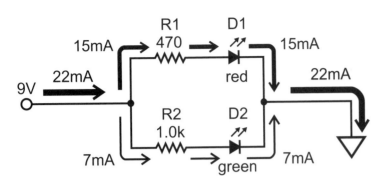

Figure 5-4.

currents will add together when they meet at the node leading to the ground, and this is in fact what we see. At the other end, the current leaving the battery is also 22mA, and it splits at the voltage supply bus and flows into the two arms of the circuit. This is shown in figure 5-4. Kirchhoff's current law is true for all the points in the circuit: at nodes where current splits, at nodes where current comes together, at components (the current entering a component must equal the current leaving it), at the voltage source (the current leaving the positive terminal equals the current coming into the negative terminal), and all of the points in between.

Kirchhoff's Laws Summarized

The essence of both of Kirchhoff's laws, for voltage and for current, is that they don't just magically appear or disappear. In a circuit with a voltage source or sources, the total of the voltage drops across the components will equal the source total. As we add more components, the voltage gets spread out through them. To return to the water analogy for a moment, if we have a water tank 200 feet above our reference level (either at sea level, at Crater Lake surface level, or at Dead Sea surface level), when we open the valve and let the water out, it can just fall straight down in a single waterfall or it can fall in several stages. The total distance it falls is going to be the same. As we add more stages or flat spots to the cascading water, the sizes of the individual waterfalls must get smaller. The total distance of the combined waterfalls must be 200 feet.

The water analogy also works for Kirchhoff's current law. Whatever the flow of water is leaving the tank, that same amount has to end up at the ground, by whatever routes it takes. We can't have 100 gallons per minute flowing out of the tank and only 50 gallons per minute flowing through the final waterfall into the lake or sea. The water can't just disappear, and it can't just appear from nowhere. If 100 gallons is flowing out of the tank every minute, 100 gallons every minute is flowing into the body of water, not more and not less. If it splits into multiple streams, the total flow of each of those streams must equal 100 gallons per minute.

Ohm's Law

In the hands-on exercise with the parallel circuit, you built a circuit with two arms. One arm was made up of a 470Ω resistor with an LED, the other of a 1.0kΩ resistor with an LED. We saw that the current in the first arm was about 15mA. The current running through the second arm was about 7mA. We may also remember that we defined resistance as the ability to restrict the flow of current: if the resistance is high, it doesn't let through very much current.

We can sum all of this up by saying that for a particular battery or other voltage source, increasing the resistance decreases the current. In the 1820s, a self-taught Bavarian scientist named Georg Simon Ohm was the first person to observe this relationship. Ohm's law is named after him. The law states that if we examine a circuit with a certain potential (voltage, symbolized by E), and a certain current (symbolized by I), and a certain resistance, (symbolized by R) that the following mathematical relationship exists between them.

Ohm's Law: $E = I \times R$ (**voltage = current × resistance**)

Ohm's law only holds true if all of the terms are expressed in their correct units. The voltage must be in volts, the current must be in amperes, and the resistance must be in ohms. If the values are in some other units (such as mV, mA, kΩ or MΩ), they must be converted to their basic units before being used in the Ohm's law equation.

We can go back to the water analogy to help illustrate Ohm's law, as shown in figure 5-5. Voltage is the pressure of the water wanting to get back to sea level. Current is the flow of water through the pipe. Resistance is a measurement of how tightly the valve is closed. A wide-open valve has very little restriction, and a closed valve has a lot of restriction. In the same way, a resistor with a lot of resistance has a lot of restriction on the current.

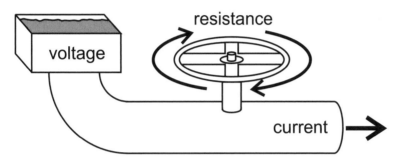

Figure 5-5.

Why Ohm's Law is Important

Ohm's law has a number of applications. One of them is that for any circuit for which we know two of the three qualities, we can calculate the third one. This is quite useful for current, since current must be measured with the ammeter inserted into the circuit. In a test circuit on a breadboard this is pretty easy, but it's usually much more problematic to determine the current flowing in the circuit of a piece of audio gear.

All right, so we can use Ohm's law to calculate resistance or current or voltage. That's all fine if we're working in a lab, or designing electronics, or whatever else one does with a degree in electrical engineering. We might wonder why audio professionals would be concerned with Ohm's law.

We need to know about it because everything in audio requires electricity. Ohm's law is at the heart of the gear we use. The purpose of the equipment is to get the right amount of current flowing to the right place at the right time. It is designed to provide the right voltage and restrict it with the right resistance. If we plug in the monitor speakers incorrectly, the power amp goes into thermal overload and shuts down because of Ohm's law. If we plug an iPod into a line-level input, it sounds terrible because of Ohm's law. If we plug a really nice ribbon mic into an expensive boutique preamp and the signal sounds muffled and dead, Ohm's law is probably responsible.

Every time we tweak the EQ, or adjust the gain, or get a signal from mic to tape, or perform digital conversion, Ohm's law is at work. It's so ubiquitous that it's almost hard to see what it's doing. We'll start with some simple examples. These may seem silly, but it's easier to see all the places where Ohm's law is in effect if we start with the easy stuff.

Using Ohm's Law

The relationship between E, I, and R can also be shown with the convenient formula circle shown in figure 5-6. The correct places of the quantities E, I, and R in the circle are in alphabetical order: E on top, I and R on the bottom. The formula circle is a quick and easy way to remind yourself of the three rearrangements of the Ohm's law equation: E = I × R, I = E / R, and R = E / I. To find these equations, simply place your thumb or finger to cover the quantity you want to know, and the relationship between the other two will be revealed. For example, if we cover the I, we see that the E is over the R, giving us I = E / R.

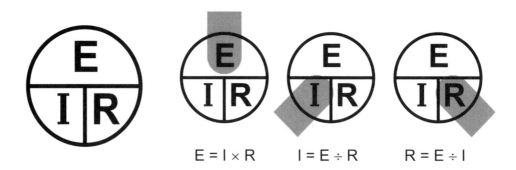

Figure 5-6.

An application of Ohm's law is for calculating the value of a resistor needed to allow through a particular current. As an example, if we are using a 9V battery with an LED that is rated as having a maximum current of 25mA, we can calculate the appropriate resistor to use.

We'll use four steps for the calculations, as shown in Appendix B.

Quantities: We list the quantities that we know, and the one quantity that we're going to solve for. If there are any prefixes, we remove them.

E = 9V
I = 25mA = 0.025A
R = ???

Equation: Identify the equation we need, and rearrange this equation so we can solve for the quantity that we don't know. Here we know the voltage (E) and the current (I), and we said above that we want to calculate the resistance (R). Ohm's law is the equation that relates voltage, current, and resistance, so this is the one we will use. We need it rearranged to get R by itself on one side:

E = I × R
R = E / I

Solve: Combine the quantities that you know with the equation, and solve the math.

R = 9V / 0.025A
R = 360Ω

Prefix: Add a prefix if one is needed. Here we don't need a prefix because 360 is already between 1 and 999.

R = 360Ω

To be on the safe side, we'd probably want a slightly lower current. Ohm's law says that if we want less current from a voltage source, we need a higher resistance, so we would use a somewhat larger resistor than this. Now we know what value to start with. A 470Ω resistor would probably be a good choice.

We'll be seeing a lot more of Ohm's law in the upcoming chapters, both directly and indirectly. One interesting application of Ohm's law is the ohmmeter function in your DMM. It supplies a known voltage between the probes and measures how much current flows through the circuit being examined. It then uses Ohm's law to calculate the resistance from the voltage and the current. This is why the ohmmeter can't be used on a working circuit where there is already voltage across the resistors and current flowing through them.

Apps and Websites

There are numerous tools on the Internet and in the app store for making Ohm's law calculations, and by all means you should be familiar with them. In class, I don't let my students use them on tests, because I believe that you need to understand how Ohm's law actually works. The only way I know of doing that is to build circuits to see it for yourself and then to get in and do the calculations by hand for a while. For example, by doing enough calculations, you will get a sense of Ohm's law and how current and voltage and resistance are related. You don't get this understanding just by plugging numbers into a calculator app. I really suggest doing both: doing the math, then using the app to check your answer.

If you want to just use an app, that's fine. I'm old-school here, because that's the way I learned. For an example of the kind of understanding I'm talking about, watch *Apollo 13*. There were situations where the engineers needed to know and feel their scientific knowledge at a gut level. Being able to key airflow and filtering capabilities into an app might have gotten them the answer, but they got there by feeling it. They didn't have pocket calculators, much less apps. All they had were slide rules and pencils, so they had to be able to get in the ballpark by really knowing their field of expertise. I can't say that my experience with electronics is close to that point, because aspects of it are still able to surprise me. It's taken me a lifetime to get here, and there is still an amazing amount to learn. On one hand, the apps and tools clear the way to think about other more interesting aspects of the material. On the other hand, I believe that they can get in the way of real understanding.

Hands-On Practice

This exercise will demonstrate the principles of Ohm's law. You'll also get some more experience with building circuits and using your DMM for measuring current, voltage, and resistance.

Materials

These are the materials you will need for these exercises.

- Digital multimeter (DMM)
- Breadboard
- 9V battery and battery clip with wires
- 560Ω resistor (banded green–blue–brown–gold)
- 1.0kΩ resistor (banded brown–black–red–gold)
- 10kΩ resistor (banded brown–black–orange–gold)
- 100Ω resistor (banded brown–black–brown–gold)
- Red LED
- Insulated solid wire (20, 22, or 24 gauge)

Measurements in the First Circuit

- Set up your breadboard with a 9V battery.

- Find a 560Ω resistor based on its color banding of green–blue–brown–gold.

- Construct the circuit shown in figure 5-7 on your breadboard. Start by using a piece of wire in place of the meter to show that the circuit works and the LED lights up. Remember that the current in the circuit must flow through the meter, so construct the circuit exactly as shown, to avoid overloading the ammeter.

- Read the current on the 20 mA scale and record it on a piece of paper.

- Remove the ammeter and reconnect the circuit so the LED is lit.

- Measure the voltage across the resistor and record it on a piece of paper.

- Disconnect the circuit by removing the wire.

- Measure the resistance of the resistor and record it on a piece of paper.

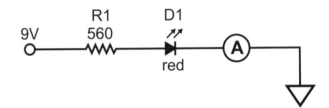

Figure 5-7.

Measurements in the Second Circuit

- Find a 1.0kΩ resistor based on its color banding of brown–black–red–gold. Be careful in locating this resistor, because the red bands on resistors can look like orange bands. Use your DMM set as an ohmmeter to check the resistor value if you need to.

- Construct the circuit shown in figure 5-8 on your breadboard, first with a wire to show that it works, then with the ammeter in its place. As before, be careful not to overload the ammeter.

- Read the current on the 20mA scale and record it on a piece of paper.

- Remove the ammeter and reconnect the circuit so the LED is lit.

- Measure the voltage across the resistor and record it on a piece of paper.

- Disconnect the circuit by removing the wire.

- Measure the resistance of the resistor and record it on a piece of paper.

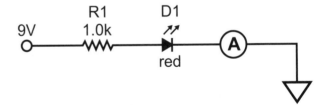

Figure 5-8.

Measurements in the Third Circuit

- Find a 10kΩ resistor based on its color banding of brown–black–orange–gold. Be careful in locating this resistor also, because the orange and red bands on resistors sometimes look similar. Use your DMM set as an ohmmeter to check the resistor value if you need to.

- Construct the circuit shown in figure 5-9 on your breadboard, first with a wire and then with the ammeter. As always, be careful not to overload the ammeter. The LED will be very dim or will not light at all.

- Read the current on the 2mA scale and record it on a piece of paper.

- Remove the ammeter and reconnect the circuit.

- Measure the voltage across the resistor and record it on a piece of paper.

- Disconnect the circuit by removing the wire.

- Measure the resistance of the resistor and record it on a piece of paper.

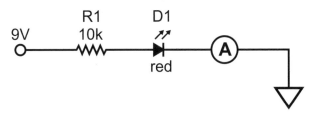

Figure 5-9.

Measurements in the Fourth Circuit

- Find a 100kΩ resistor based on its color banding of brown–black–yellow–gold.

- Construct the circuit shown in figure 5-10 on your breadboard, again being careful not to overload the ammeter. The LED will not light, so don't worry when it doesn't seem to be working.

- Read the current on the 200μA scale (if available) and record it on a piece of paper. If the meter doesn't have a 200μA scale, just use the lowest setting.

- Remove the ammeter and reconnect the circuit.

- Measure the voltage across the resistor and record it on a piece of paper.

- Disconnect the circuit by removing the wire.

- Measure the resistance of the resistor and record it on a piece of paper.

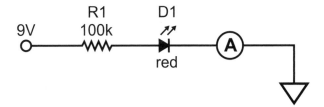

Figure 5-10.

Observations from the Measurements

Looking at the measurements you've just made, you've seen that the current is reduced in each successive circuit. This is because the resistance in each circuit has been increased each time. We see this change in current because of Ohm's law. The battery voltage is not changing, so the voltage in the equation is not changing. The result is that when the resistance is increased, the current decreases. This makes sense, because resistance is a measurement of how much restriction is being placed on the current.

Ohm's Law Calculations

This exercise has demonstrated Ohm's law for you by having you measure the current in a circuit where you are keeping the voltage supply constant and steadily increasing the resistance. As the resistance values were raised, the current was observed to decrease. This set of problems will give you experience with calculating voltage, current, and resistance using Ohm's law.

Set 1

Use Ohm's law to calculate the current in the circuit shown in figure 5-11. Use the four steps as shown in the chapter. The answer is given at the end of the exercise.

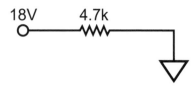

Figure 5-11.

Set 2

Use Ohm's law to calculate the current in the circuit shown in figure 5-12. Use the four steps as shown in the chapter. The answer is given at the end of the exercise.

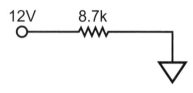

Figure 5-12.

Set 3

Use Ohm's law to calculate the resistance in the circuit shown in figure 5-13. Use the four steps as shown in the chapter. The answer is given at the end of the exercise.

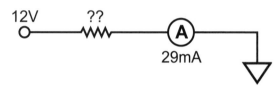

Figure 5-13.

Set 4

Use Ohm's law to calculate the resistance in the circuit shown in figure 5-14. Use the four steps as shown in the chapter. The answer is given at the end of the exercise.

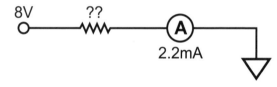

Figure 5-14.

Set 5

Use Ohm's law to calculate the voltage in the circuit shown in figure 5-15. Use the four steps as shown in the chapter. The answer is given at the end of the exercise.

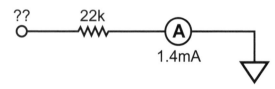

Figure 5-15.

Set 6

Use Ohm's law to calculate the voltage in the circuit shown in figure 5-16. Use the four steps as shown in the chapter. The answer is given at the end of the exercise.

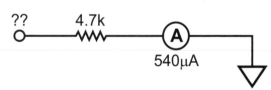

Figure 5-16.

Answers

Set 1

Quantities
E = 18V
R = 4.7kΩ = 4700Ω
I = ???
Equation
I = E / R
Solve
I = 18V / 4700Ω
I = 0.00382979A
Prefix
I = 3.8mA

Set 2

Quantities
E = 12V
R = 8.7kΩ = 8700Ω
I = ???
Equation
I = E / R
Solve
I = 12V / 8700Ω
I = 0.0013793
Prefix
I = 1.4mA

Set 3

Quantities
E = 12V
I = 29mA = 0.029A
R = ???
Equation
R = E / I
Solve
R = 12V / 0.029A
R = 413.793Ω
Prefix
R = 414Ω

Set 4

Quantities
E = 8V
I = 2.2mA = 0.0022A
R = ???
Equation
R = E / I
Solve
R = 8V / 0.0022A
R = 3636.3636Ω
Prefix
R = 3.6kΩ

Set 5

Quantities
I = 1.4mA = 0.0014A
R = 22kΩ = 22,000Ω
E = ???
Equation
E = I × R
Solve
E = 0.0014A × 22,000Ω
E = 30.8V
Prefix
E = 30.8V

Set 6

Quantities
I = 540µA = 0.00054A
R = 4.7kΩ = 4,700Ω
E = ???
Equation
E = I × R
Solve
E = 0.00054A ×4,700Ω
E = 2.538V
Prefix
E = 2.5V

6

AC Voltage

Goal: When you have completed this chapter, you will have gained some insights into how AC voltage and DC voltage are different, and in what ways they are similar.

Objectives

- Identify the difference between AC voltage and DC voltage.
- Recognize the layout of an AC wall outlet.
- Test the AC outlet safely with a DMM.
- Identify the relationship between DC voltage, AC peak-to-peak voltage, and RMS voltage.

AC Voltage

So far all the voltage sources we have talked about have provided direct current (DC), which is produced by batteries and DC power supplies. All the electronic equipment you use in a studio uses DC voltage. As the name implies by including the word *direct*, DC voltage remains constant, either positive or negative: +9V is a typical supply voltage for effects pedals, preamps, and the like. Power amps usually use a higher voltage, but it is still a consistent voltage such as +24V or +40V. Tube equipment uses very high voltages, usually around 350V. Another example of DC voltage is the +48V potential of phantom power used for supplying condenser microphones.

AC voltage fluctuates above and below the ground as time passes. One example is the voltage that is found at the wall socket. This is called *line voltage*, *wall voltage*, or *mains voltage*. In North America and Japan, this voltage has a level of 110V to 120V. Most other countries use a voltage level of 230V. In North America, the changing voltage has a frequency of 60 hertz (60Hz). The cycle time of one full wave repetition (from ground to maximum positive to ground to maximum negative to ground) is 1/60 of a second. The hum we hear from problematic amplifiers is called "60 cycle hum," because the older term for hertz was "cycles per second." Europe uses 50Hz line voltage, as does much of Asia.

The shape of the current fluctuations of AC voltage from the wall outlet is a sine wave, which is a pure tone of a single frequency with no other frequencies added in. You are probably already familiar with the shape of a sine wave, which rises and falls smoothly as time passes. Four cycles of an AC line voltage wave are shown in figure 6-1.

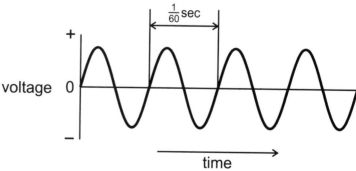

Figure 6-1.

It should be noted that the term "line" gets applied to two kinds of AC voltage. One is the line voltage of the AC supply from the wall socket. The other is the line-level signal used in audio gear. While these are both AC, they are distinctly different. For this reason, I will always refer to the AC voltage at the wall socket as *supply line voltage* to distinguish it from line-level audio signal. Other references don't make this distinction, so it's vital to be aware of which voltage is being discussed.

AC Is Alternating, DC Is Direct

Up to now we've been maintaining very specific terminology when talking about voltage (the potential) compared with when we're talking about current (the movement of electrons). When discussing AC voltage and DC voltage, we seem to be mixing up these terms: if we spell them out, AC voltage becomes "alternating current voltage" and DC voltage becomes "direct current voltage." This is not really what we mean. AC voltage is simply voltage that alternates between positive and negative, whether there is current flowing or not. Similarly, DC voltage is just voltage that stays positive all the time or negative all the time. If we were to be very literal, we would be saying that a battery is a source of "direct voltage," and the wall outlet is a source of "alternating voltage." In both cases, the potential is sitting there waiting to be used, but current doesn't flow until we provide a circuit for it to travel. "AC" has come to mean "alternating," and "DC" has come to mean "direct."

What's Happening in the Conductor

In a circuit with a DC source, the electrons are all flowing in the same direction as long as the circuit is energized: the instant the circuit is completed, there is a wave of "pressure" from the source that moves through the conductor at about one-third the speed of light (although the electrons themselves are actually moving through the conductor at a very slow rate of a few centimeters per hour). We can think of the positive DC voltage source as constantly pushing current into the circuit and toward the signal ground: "push, push, push." Since this is a constant, steady push, a better word might be "puuuuuuuuuuush." A negative DC voltage source is constantly trying to pull current out of the circuit, away from the signal ground: "pull, pull, pull," or even better, "puuuuuuuuuuuull." These movements are shown in the top two panels of figure 6-2.

The last panel of figure 6-2 shows the situation with an AC source. In an AC circuit, the particles spend half the cycle moving in one direction and the other half moving in the other direction. Their total movement over time is zero. Each time the potential from the source changes from positive to negative, the "pressure" from the source changes to "suction." These waves of alternating pressure and suction cause the electrons in the entire length of the conductor to move back and forth in place. They still do work, because they're moving, but they don't actually go anywhere in the process. We can think of the AC voltage source as pushing and pulling the electrons: "push, pull, push, pull, push, pull."

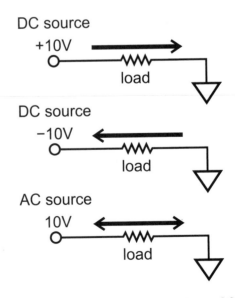

Figure 6-2.

Impedance

Speakers, microphones, amplifiers, and many other kinds of other electronic gear and components each have an inherent quality called *impedance*, and every time we combine them together, their impedances will interact with each other. The symbol for impedance is **Z**. This property is measured in ohms (Ω), just like resistance. Impedance and resistance are similar yet different. They can both be described as having the ability to restrict the flow of current. The difference is that the resistance stays the same regardless of whether the voltage is DC or any frequency of AC, while impedance acts differently with AC and DC. Specifically, impedance differs depending on what frequency of AC is being supplied.

Impedance is actually the combination of two quantities: resistance (R) and *reactance* (X). We already know about resistance. Reactance in turn is made up of two types: *inductive reactance* (X_L), which is the reactance from energy being stored in a magnetic field (in an inductor); and *capacitive reactance* (X_C), which is reactance from energy stored in an electric field (in a capacitor). We haven't learned about capacitors and inductors yet, and this is not the right place to introduce them. We're just going to say that they can store energy, and that any time we have a capacitor or an inductor involved, we're introducing an element into the circuit that exhibits a resistance that varies with frequency.

The math gets a bit tricky here, with imaginary numbers and vector addition, but we're not going to worry about it too much. These two reactances are each 90° out of phase with the input signal, one leading and one lagging, and this adds to the complexity of calculating the total reactance and thus the impedance. The bottom line for us is that impedance is resistance that changes when we change the frequency. We can't measure impedance directly with a meter the way we can resistance. How impedance works is quite a difficult concept to grasp fully, so don't worry if it takes a while to find the right explanation for you. We'll talk further about impedance in the later chapter on voltage dividers, because a voltage divider is a very useful model to help understand how the output impedance of an audio source (such as a microphone) interacts with the input device that it is plugged into (such as a preamp).

Measuring AC Voltage

Before we grab the DMM and set to work measuring AC voltage, we need to consider what the waves of this voltage look like and how their shape will affect their measurement.

If we take an average of the voltage over time, it will be zero. For every push, there is an equal pull an instant later, and the electron ends up back at the same place. The result is that the average of AC voltage is zero, so it's not a useful way of talking about the wave. We need a different solution to express how strong the potential of the wave is.

We can use an oscilloscope to look at the shape of the wave that is changing as time passes. If we look at our AC voltage with an oscilloscope, we can see the voltage changing up and down above and below the ground level. An example is shown in figure 6-3.

We can look at the screen and see how big the fluctuations are by measuring the height of the wave on the display. The distance from the

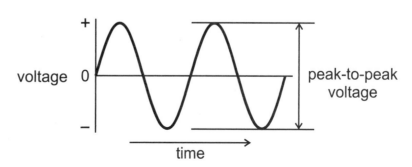

Figure 6-3.

top of the positive peak to the bottom of the negative peak is called the *peak-to-peak voltage*. This is the easiest quantity to measure on the oscilloscope, because we can see the top and bottom of the wave on the screen.

The DMM doesn't come equipped with an oscilloscope for looking at the peak-to-peak measurement. Even if it did, we are usually less concerned with the peak-to-peak value of an AC wave and more concerned with how much work it can do. Whether we are measuring an AC voltage source or a DC source, we want to know how much potential the source has to do work. When measuring an AC source, it's not useful to average the voltage over time, because the top half of the wave and the bottom half of the wave are equal and add up to zero. It's also not useful to try to display the voltage at a particular moment in time, because it is always changing. What we need is to display how much potential energy is available to use over a period of time.

The official name for this measurement is the *root-mean-squared calculation*, or RMS for short. There are two ways for us to determine the RMS voltage from the peak-to-peak voltage. One way is to use calculus. The other way is to use chocolate. There are a lot of calculus books that show how to do the RMS calculation, but I believe this is the first book to use chocolate to demonstrate how to calculate RMS voltage from peak-to-peak voltage.

Let us say that we have an AC wave that is 6V peak-to-peak. We can cut a slab of chocolate 1" thick into a model of this wave measuring a total of 6" tall and 5" long. (This 5" dimension could be any number, but I chose 5" to make it easier to keep track of.) This chocolate bar is shown in figure 6-4.

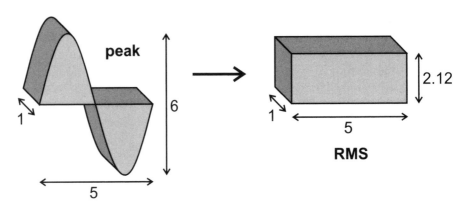

Figure 6-4.

We can then melt the chocolate and pour it into a mold that is 1" thick and 5" long. If we didn't lose any chocolate along the way (as might happen with melted chocolate), we would end up with a bar 1" thick (just like the AC wave bar), 5" long (also just like the AC wave bar), and just a little more than 2" tall.

What this melted-chocolate method shows us is the same thing that calculus shows us: the area under the curve. Calculus tells us exactly the relationship between the total height of the wave and its area: when we melt it down, it becomes 35.4% of the total height. For our 6V wave, the voltage is reduced to 2.12V, as represented by the 2" chocolate bar.

This RMS voltage is what is being measured by the DMM: the melted-down, how-much-work-can-it-do measurement of AC voltage. The RMS value represents what DC voltage would do the same amount of work as the AC voltage being measured. The relationships between the peak-to-peak and RMS measurements of the voltage at the electrical outlet are shown in figure 6-5. The DMM reads 120VAC. For the same wave, the peak-to-peak voltage is 340V.

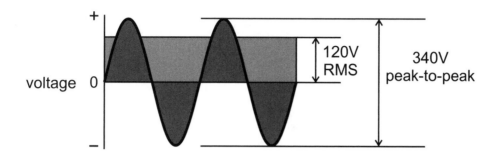

Figure 6-5.

AC Voltage Is Always a Positive Number

We said that the electrons in the conductor are moving back and forth: push, pull, push, pull. When we turn off the voltage source and the electrons stop moving, they have not moved anywhere. Their average movement is zero.

Whether it's a peak-to-peak voltage or an RMS voltage, AC voltage is expressed as a positive number. A peak-to-peak voltage value is an expression of a distance—namely, the distance between the peaks of the wave. That distance is always going to be zero or more, because any distance that isn't zero is a positive value. An RMS voltage is an expression of the area under the curve. The smallest area available is zero. Any area that isn't zero is a positive value.

The end result is that AC voltage is always either a positive number or zero. DC voltage can be positive, zero, or negative.

Other Types of AC Voltage

The 120V AC line voltage from the wall socket is not the only AC voltage to be aware of. Audio signals also fluctuate above and below the ground as time passes. This wave could be any shape, such as sine wave, triangle wave, white noise, a vocalist singing, and so on. These are all AC voltages, because they are fluctuating above and below the ground. The source could be a microphone, guitar pickup, turntable cartridge, analog magnetic tape head, or digital-to-analog converter (such as in a CD player, sound card, digital recorder, etc.).

One example of an AC voltage source is a dynamic microphone. At rest, it generates no voltage or current. When a sound wave in the air hits it and makes the diaphragm move, the coil moves relative to the fixed magnet, which generates a positive voltage. When the sound pressure is released, the coil moves back out, generating a negative voltage.

The Electrical Outlet

A grounded *electrical outlet* has three holes or slots, as shown in figure 6-6. (Older buildings sometimes still have outlets with just two slots and no ground.) Each has a particular connection to the electrical supply.

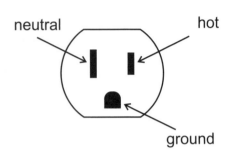

Figure 6-6.

The half-round hole at the bottom is the earth ground. This contact is connected to a big copper rod or pipe stuck into the earth. This connection to the earth has a low resistance. In the event of an equipment malfunction, current from the equipment will prefer to go to the earth by way of the ground wire instead of by way of some other path (such as through you). The standard wire insulation color for earth ground in electrical wiring is green, or the wire may be bare ("green for ground").

The smaller slot is the "hot" electrical supply. This is the dangerous one, providing 120V of alternating current, from wherever the electricity is generated: a turbine at a hydroelectric plant turned by water flowing over a dam, steam from a nuclear plant or a coal or oil plant, or a gasoline generator. The standard wire color for hot is black. This is the wire that carries the current to the *load*, which is the device doing the the work.

The larger slot is the "neutral." This is the path that connects the other side of the load (the resistor, the motor, the transformer, the power amp) back to the source. Neutral is the return path for the current that is moving in the wiring when work is being done on the load. The standard wire color for neutral is white.

Depending on the local electrical code and the equipment, neutral may or may not be connected to the earth ground. In any case, it should be at zero potential—i.e., at the same potential as the ground. Just because it is at the same potential as the ground doesn't mean it is safe! Remember that since neutral is the return path for the current, all the current that is flowing through the load is also flowing through the neutral. If you interrupt the return path and put your own body in its place, you become the return path, and whatever current is flowing through the load is now flowing through you! I've done this by accident a few times, and I can vouch that it was not an enjoyable experience.

The reason that the larger slot is the neutral is that if there is a short in the circuit between the hot and the neutral, current will flow without restriction and the wire will catch fire or melt. The terminal that melts is likely to be the smaller one, so if the neutral terminal is smaller and melts

away, the rest of the circuit remains charged from the hot terminal. This situation could easily result in a deadly shock if a person touched the charged circuit. Consequently the hot terminal is made smaller than the neutral, so that it melts or burns out first, and the rest of the circuit is not charged. In theory, the circuit breaker or fuse protecting the circuit should detect the excessive current and turn off the circuit automatically, but I've personally experienced the catastrophic failure of a device that caused the entire electrical outlet to melt like one of Salvadore Dali's clocks, as shown in figure 6-7.

Figure 6-7.

The GFCI

One way to protect yourself from electrical shock is to use a GFCI outlet. GFCI stands for *Ground Fault Circuit Interrupter*. What it does is exactly what the name says: if it detects a ground fault, it interrupts the circuit. A ground fault occurs when current finds a path to ground, and usually this path is through a human.

We've all seen GFCI outlets: they are the ones with the pair of buttons in the middle of the outlet, as shown in figure 6-8. Occasionally the reset button pops out and the outlet turns itself off. These outlets are now required by the National Electrical Code in bathrooms, kitchens, garages, and outdoors. These are places where a person, water, and electricity are likely to come together, with the person getting the worst of the interaction.

The GFCI works on Kirchhoff's current law, which we can sum up with the statement that current in equals current out. The actual protection circuitry in a GFCI does not need to be covered here. What is useful to know is that in normal use, the amount of current flowing out to the thing that we have plugged in must equal the current flowing back into the socket. This is shown in the first panel of figure 6-9.

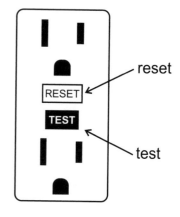

Figure 6-8.

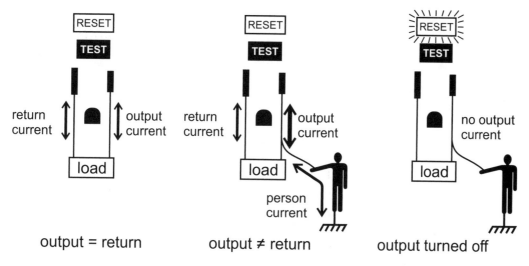

Figure 6-9.

The outlet is a source of AC, with the current moving back and forth in the conductor and doing the work we want it to do in the load (amplifier, console, battery charger, blow dryer, or whatever it is that we have plugged in). When current is flowing out of the hot connection, it is also moving into the neutral connection. During the negative half of the wave, current flows out of the neutral connection and gets pulled into the hot connection. At all times, the amount of current flowing out must be equal to the amount returning to the outlet. The GFCI is constantly monitoring the flow of current to check that the current going out and the current coming back in are at the same level.

If someone comes along and touches a live wire attached to the GFCI (either the hot or the neutral), that person becomes a path to ground, and current flows through the person. This is shown in the middle panel of figure 6-9. Now there is current flowing along two paths: through the load and through the person. The GFCI notices right away that the current going out is greater than the current coming back in, and it trips its circuit breaker. This turns off the current flow and pops out the reset button, as shown in the third panel. It does this so quickly (less than 0.1 second) that the person who was about to get an electrical shock doesn't even notice.

GFCIs are available for mounting as wall outlets, as portable modules that plug into standard outlets, as parts of extension cords, and as circuit breakers for permanent mounting to protect an entire circuit. Permanent installations should be tested every month with the test button. Portable GFCIs should be tested every time they are plugged in. If the reset button doesn't pop out when the test is pressed, the GFCI should be discarded and replaced.

I mention GFCIs for two reasons. The first is that a GFCI is the only safe place to start your experience with measuring AC line voltage with your DMM. After sufficient practice, you can go out and measure AC supply line voltage at any outlet safely, but it's best to start with the GFCI as your safety net. The second reason is that I strongly recommend including a GFCI in your electronics workbench. A GFCI will not provide full protection from electrical shock under every circumstance of testing electronic gear, but it is an effective tool to help protect yourself.

Measuring AC Supply Line Voltage with the DMM

To measure AC supply line voltage at the outlet, your DMM needs to be set to the AC voltage range. The setting may be marked with **VAC, ACV,** or **V** with a squiggly line to indicate alternating current. The probes need to be plugged into the meter in the correct places for measuring voltage. For your safety and also for the preservation of your meter, if there is any doubt about how much voltage is present, you should use the highest setting to start with. In North America, we can be fairly sure that our local electrical utility is providing something at least reasonably close to 120VAC, so we can use the 200V setting. This is shown in the photo in the hands-on exercise.

The consequences of using the incorrect setting for measuring supply line voltage range from mild to deadly. If you try to measure the voltage at the outlet using the DC voltage setting, nothing much will happen, because the meter is already expecting a voltage. On the setting for current, the fuse would blow, or the Polyswitch would likely melt. On the settings for resistance and continuity, the high potential could totally destroy the insides of the meter. I once had a student who borrowed a DMM from his friend and set it incorrectly to test an electrical outlet on the final exam. The student escaped unharmed, but the meter was rendered useless.

Hands-On Practice

Materials

These are the materials you will need for these exercises.

- Digital multimeter (DMM)
- GFCI outlet
- Optional: dead piece of audio gear

This exercise could kill you or injure you. It's the grown-up version of sticking a paper clip in the electrical socket, so it's reasonable to be concerned about how safe it is. With appropriate caution, you will be able to become comfortable with measuring supply line voltage without a second thought. In the meantime, follow each step very carefully and deliberately. I strongly recommend practicing with a GFCI-protected outlet until you are very comfortable with the procedure. At some later time, you will probably have to measure voltage at an unprotected outlet, but it's better to learn in a safer environment.

- This exercise must be performed using a GFCI-protected outlet.

- Before starting, test the GFCI by pressing the test button. The reset button will pop out if the GFCI is working properly.

- Press the reset button.

- Set your meter to read AC voltage. This is the voltage setting with the squiggle, which indicates AC. If your red probe is movable, check that it is in the socket labeled **V**. The black probe should be in the common socket as usual. This setting is shown in figure 6-10.

- Double-check that your meter is on the correct setting and that the red probe is in the V socket. If the probe is in the wrong socket or on the wrong scale, you can destroy your meter by trying to measure supply line voltage with it.

- As always, but in particular when you are measuring high-voltage AC, remember to keep your fingers away from the metal parts of the probes.

- You will be measuring the voltage at the neutral and hot slots of the outlet. It is also correct to measure at the hot and ground slots, but it can be dangerous, because the slot for the ground prong of the 3-prong is so much larger than the DMM probe. It's safer to connect to the narrower neutral slot first, then to the hot. To make contact with the inside of the ground slot, you need to turn the probe at a slant to touch the inner side of the slot.

Figure 6-10.

- Insert the black probe into the large slot (neutral) of the AC socket.

- Insert the red probe into the smaller slot (hot) of the socket. This is shown in figure 6-11.

- Read the measurement and record it on your piece of paper. This reading should be somewhere around 110V. If you are not seeing this voltage, try moving the probes around in the slots to ensure better contact.

- Remove the red probe.

- Remove the black probe.

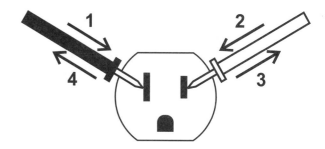

Figure 6-11.

Reversing the Probes

When measuring AC voltage, it doesn't matter whether the black probe or the red probe is the one placed in the neutral slot. We use the black probe by habit.

- Insert the red probe into the large slot (neutral) of the AC socket.

- Insert the black probe into the smaller slot (hot) of the socket.

- Take the reading and record it on your piece of paper.

- Remove the black probe.
- Remove the red probe.
- Turn off your meter.
- Compare your reading with what you got when you measured earlier with the probes in the correct positions. It should be identical.

Optional: Dead Piece of Audio Gear

If you have a piece of audio gear available to work on and it runs on supply line voltage (with a standard 3-prong or 2-prong plug), you can examine it for where the AC supply goes inside and use your DMM to test it.

- Do not plug the unit in yet.
- Open the box and look inside for where the AC supply comes in. You should see a white wire, a green wire, and a black or blue wire. The white wire is the neutral, the return path. The green wire is the earth ground if it has one. The black wire is the hot.
- Trace these wires to where they go next, and see if you can make contact with your DMM.
- Plug the unit into a GFCI.
- Turn it on if it has a power switch.
- Examine the contact points inside the box with your DMM. Your black DMM probe should contact where the white wire goes, and your red DMM probe should contact where the black wire goes.
- You should get a reading of 120V, just as you did from the outlet.

This exercise has given you the experience of safely measuring AC voltage with your DMM. You will find this a useful skill in the studio and in live sound.

7

The Resistor Code

Goal: When you have completed this chapter, you will be able to interpret the pattern of colored bands on any resistor to determine its value and tolerance.

Objectives

- Interpret the 4-band color coding of a resistor.
- Interpret the 5-band color coding of a resistor.
- Use the DMM to confirm the value of a resistor.
- Use the color code to determine the 4-band and 5-band patterns for any desired resistor.
- Recognize SMT resistors.

The Resistor Code

Resistors, like all electronic components, are made in a wide variety of shapes and sizes. The most usual type is a small cylinder. These cylindrical resistors are color coded for easy identification, showing us the value of the component. The color codes are shown in table 7-1. There are several reasons for the color code: once you know which direction to read the bands, it's always the right way up; and it's easier to print on the resistor than tiny numbers. Modern miniature resistors do have tiny numbers printed on them to identify them, thanks to great strides in printing since the code was first implemented.

Using the Resistor Code with 4-Band Resistors

Some resistors, such as the one shown in figure 7-1, have four bands on them. In order, they are designated as the first digit, second digit, multiplier, and tolerance.

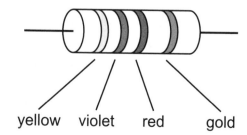

yellow   violet   red   gold

Figure 7-1.

Resistor band color codes			
digit codes		tolerance codes	
color	digit	color	tolerance
black	0	brown	±1%
brown	1	red	±2%
red	2	gold	±5%
orange	3	silver	±10 %
yellow	4	no band	±20 %
green	5		
blue	6		
violet	7	multiplier codes	
gray	8	gold	÷10
white	9	silver	÷100

Table 7-1.

In our example resistor, the first digit band (yellow) stands for 4. The second digit band (violet) stands for 7. These bands don't need any interpretation except to just read them as numbers.

The third band, the multiplier, tells us the number of zeroes to add. In this case, red stands for 2, so we add two zeroes to the end of our 4–7 sequence from above: 4–7–00. This gives us 4700Ω. This is usually written as 4.7kΩ.

The tolerance tells us how much variation is allowed from the listed value. A 4-band resistor has a gold band that stands for ±5%. Older gear may have resistors with a silver band, standing for ±10%, but you won't find this very often any more. On really old equipment you might see carbon composition resistors with just three bands. The lack of a fourth band indicates a 20% tolerance. These resistors were common once, but manufacturing methods have since become more precise. As a result, the price of 5% resistors dropped low enough that there was no longer any demand for 10% and 20% resistors.

When you first measured the resistance of several resistors of the same value, you certainly saw a little bit of variation. Of the 100Ω resistors you measured, none of them measured at exactly 100.0Ω. Instead, they measured 100.6Ω, 99.5Ω, and so on. They all fell within a certain range, which is defined by the tolerance band of the resistor. A 5% tolerance on a 100Ω resistor means that the actual value of the resistor can vary by as much as 5Ω either way, so the measured value could be anything between 95Ω and 105Ω.

Now let's look at the example resistor again. It is banded yellow–violet–red–gold. If we turned it around, it would read gold–red–violet–yellow. How do you know in which direction to read the bands? The answer lies in the gold tolerance band. This band is always the last band on the resistor, so you start reading at the other end. There is supposed to be a larger gap between the tolerance band and the others, but the difference is often so small as to be undetectable.

Now that we know which direction to read it in, the color code gives us the following: 4–7–2–5%. We interpret this 4–7–2–5% as 4700Ω±5%, which becomes 4.7kΩ±5% when we add the standard prefix.

That's all fine for identifying a resistor from its colors when we find one. We will also need to be able to start with a desired resistance and determine what color band pattern to look for. There is an easy sequence of steps to follow to figure out the color bands for any desired resistor. For example, we might need to know the color band pattern for a 180kΩ resistor. Unless we say specifically otherwise, we'll assume we want a resistor with a 5% tolerance.

It is worth mentioning that when I'm teaching this procedure in a class, there's a strong tendency to jump right in and start calling out the color codes for particular numbers. It's easy for the first two numbers, and then people get a little frustrated because it's not immediately apparent what the multiplier should be. I really recommend that you follow the slightly longer procedure that I show here until you start to become familiar with the codes. From now on, all the resistors in the hands-on section will just have their values listed without their color bands. This will give you practice with learning the color codes. It's a longer road initially, but it gets easier the more you practice.

Try this for the 180kΩ resistor with 5% tolerance.

Step 1. Write out the desired value in full, without the prefix, but with the units (Ω) and the tolerance. For our example, this is 180,000Ω±5%.

Step 2. Write out the first digit and the second digit, then write out the number that will represent the zeroes that remain, and finally, the tolerance. In our example, this is 1–8–4–5%. The "4" is for the four zeroes that are left over after we take off the first two digits.

Step 3. Change this sequence of numbers to a sequence of colors: 1 = brown, 8 = gray, 4 = yellow, 5% = gold. Now you know that you need to find a resistor that is banded brown–gray–yellow–gold. Even when you know the colors you are looking for, this is not easy when you are first starting out. Good light and a magnifying glass or a Coddington magnifier are tools that can help in your search.

Here's a second example: 22Ω resistor with 5% tolerance.

Step 1. Write out, with no prefix: 22Ω±5%.

Step 2. Write out the first digit and the second digit, then the number that will represent the zeroes that remain, and finally the tolerance. For this resistor, this is 2–2–0–5%. The "0" here is for the number of zeroes that remain after we write "2" for the first digit and "2" for the second—namely, that there are no zeroes at all.

Step 3. Change this sequence of numbers to a sequence of colors: 2 = red, 2 = red, 0 = black, 5% = gold. The resistor you want is banded red–red–black–gold.

If the resistor is even smaller than this, we need "fewer than zero" zeroes. Here's our procedure on a 3.6Ω resistor with 5% tolerance.

Step 1. Write out, with no prefix: 3.6Ω±5%.

Step 2. Write out the first digit and the second digit, then the number that will represent the zeroes that remain, and finally the tolerance. For this resistor, this is 3–6– . . . and now what? Even with no zeroes, we need to divide by 10 to get 3.6 from the 3–6– represented by the first two bands. Fortunately, the resistor code has a way of dealing with this rare situation: a gold band for the multiplier stands for "divide by 10." This means that what a gold band represents depends on whether it is in the multiplier position or the tolerance position. When it's the last band, it's a tolerance of ±5%, and when it's the second to last band, it stands for "divide by 10," which is the same as "multiply by 0.1." This gold multiplier band is shown in the "multiplier codes" box of table 7-1. With this knowledge we can write out the parts of this resistor's code: 3–6–"divide by 10"–5%.

Step 3. Change this sequence of numbers to a sequence of colors: 3 = orange, 6 = blue, "divide by 10" = gold, 5% = gold. The resistor you want is banded orange–blue–gold– gold. The two gold bands have different meanings based on their positions.

This example shows that the multiplier band can have a gold band standing for "divide by 10." The table also shows that a silver band in this position stands for "divide by 100."

Using the Resistor Code with 5-Band Resistors

Modern metal film resistors have five bands. In order, they designate the first digit, second digit, third digit, multiplier, and tolerance. The code is identical to the 4-band code, except that an additional digit has been added. The tolerance rating of these resistors is usually 1%, shown by a brown band. There should be an extra gap between the tolerance band and the others, but it's very hard to see. An example is shown in figure 7-2, with a color code of brown-black-green-red-brown.

The first digit band (brown) stands for 1, the second digit band (black) stands for 0, and the third digit band (green) stands for 5. This is just the same as for the 4-band resistor, but with one additional digit. Together these three bands give us 1–0–5.

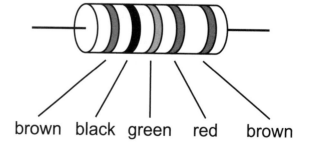

brown black green red brown

Figure 7-2.

This time the fourth band is the multiplier, and it's red, which stands for 2. We add two zeroes to the end of our existing 1–0–5 sequence from the digit bands, for 1–0–5–00. Written all together, this is 10,500Ω. This is usually written as 10.5kΩ.

The last band is the tolerance band. In this case, brown stands for 1%, just as brown stands for 1 in the digit code. The final interpretation is 10.5kΩ±1%.

It is sometimes hard to see the larger gap between the tolerance band and the others, so that a resistor with a brown band at each end can be confusing. Measure the actual resistance with a DMM if there is any doubt.

Memorizing the Resistor Code

It really helps to memorize the color code so that you can read resistors easily and without thinking about it too much. There are also smartphone apps for reading band colors, but to my mind it's still easier to just memorize them. Here are some mnemonics to help you remember the order of the colors.

- Black Beetles Running On Your Garden Bring Very Good Weather.
- Better Be Ready, Or Your Great Big Venture Goes West.
- Blackberry Brandy Rots Our Young Guts But Vodka Goes Well.
- Bold Beautiful Roses Occupy Your Garden But Violets Grow Wild.

Figure 7-3.

Another way to remember the colors in order is that most of them are the colors of the rainbow that we all learned in art class, symbolized by the somewhat silly name Roy G. Biv: red, orange, yellow, green, blue, indigo, violet. The resistor code doesn't include indigo, but it has the rest: red = 2, orange = 3, yellow = 4, green = 5, blue = 6, violet = 7. That just leaves black, brown, gray, and white. All of these are in alphabetical order: black = 0 and brown = 1, and then the rainbow colors, and then gray = 8 and white = 9. When you have the colors memorized in order, don't forget that black = 0. If you start off with black representing 1, all of your band colors will be wrong.

SMT Resistors

Many resistors in newly manufactured equipment are very small and are marked with tiny numbers instead of bands. These resistors are in Surface Mount Technology (SMT) format. These components are highly miniaturized, to make their footprints smaller. Some SMT resistors are shown in figure 7-3. These resistors are marked with either three or four numbers. The 3-number code is for digit-digit-multiplier, and the 4-number code is for digit-digit-digit-multiplier. These resistors are not marked with the tolerance, which is established by the manufacturer. The end of a resistor lead is shown at the side of the photo for scale.

Hands-On Practice

This exercise will give you practical experience with reading the resistor color code and confirming your interpretations.

Materials

These are the materials you will need for these exercises.

- Digital multimeter (DMM)
- Assortment of resistors from your exercise list
- Optional: grab bag of resistors
- Optional: dead piece of audio gear

Determining Resistor Band Colors

Through the rest of the book and out in the real world, you will need to be able to identify resistors from their band colors. To get ready for that, you should practice on some sample resistor values. I'll walk you through the first few, and then you can practice on your own. The answers are found at the very end of the exercise so you can check your work. There are a variety of 4-band and 5-band resistors.

- First example: Suppose you need a 220Ω resistor, with a 5% tolerance (4-band resistor).
    Step 1. 220Ω±5%
    Step 2. 2–2–1–5%
    Step 3. red–red–brown–gold

- Second example: Suppose you need a 332Ω resistor, with a 1% tolerance (5-band resistor).
    Step 1. 332Ω±1%
    Step 2. 3–3–2–0–1%
    Step 3. orange–orange–red–black–brown

- Third example: Suppose you need a 47kΩ resistor, with a 5% tolerance (4-band resistor).
    Step 1. 47,000Ω±5%
    Step 2. 4–7–3–5%
    Step 3. yellow–violet–orange–gold

- Fourth example: Suppose you need a 15.8kΩ resistor, with a 1% tolerance (5-band resistor).
    See end of the exercise for the answer.

- Fifth example: Suppose you need a 2.2MΩ resistor, with a 5% tolerance (4-band resistor).
    See end of the exercise for the answer.

- Sixth example: Suppose you need a 6.49kΩ resistor, with a 1% tolerance (5-band resistor).
    See end of the exercise for the answer.

- Seventh example: Suppose you need a 750kΩ resistor, with a 5% tolerance (4-band resistor).
    See end of the exercise for the answer.

- Eighth example: Suppose you need a 57.6Ω resistor, with a 1% tolerance (5-band resistor).
    See end of the exercise for the answer.

Unknown Resistor Identification

- Identify the value and tolerance of the resistor #1 shown in figure 7-4. The answer is given at the end of the exercise.

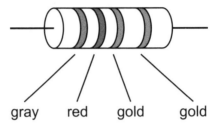

gray   red   gold   gold

Figure 7-4.

- Identify the value and tolerance of the resistor #2 shown in figure 7-5. The answer is given at the end of the exercise.

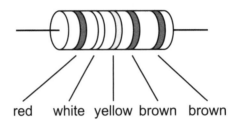

red   white   yellow   brown   brown

Figure 7-5.

- Identify the value and tolerance of the resistor #3 shown in figure 7-6. The answer is given at the end of the exercise.

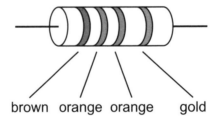

brown   orange   orange   gold

Figure 7-6.

- Identify the value and tolerance of the resistor #4 shown in figure 7-7. The answer is given at the end of the exercise.

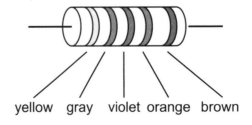

yellow   gray   violet   orange   brown

Figure 7-7.

- Identify the value and tolerance of the resistor #5 shown in figure 7-8. The answer is given at the end of the exercise.

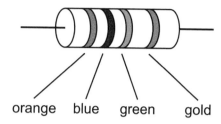

orange   blue   green   gold

Figure 7-8.

- Identify the value and tolerance of the resistor #6 shown in figure 7-9. The answer is given at the end of the exercise.

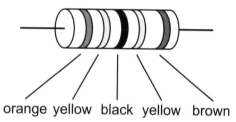

orange yellow black yellow brown

Figure 7-9.

Unknown Resistor Identification with Real Resistors

- For each of the resistors in your exercise list, identify the value of the resistor from the pattern of colored bands. Confirm your identification by measuring the value of each resistor with your DMM.

Grab Bag Measurements and Dead Gear Measurements

- If you have a grab bag of resistors, practice reading their band patterns.
- With each resistor that you identify by color code, use your DMM to confirm that you figured it out correctly. This will give you experience with both reading the bands and using your DMM.
- Earlier you started with the DMM in the middle range for measuring an unknown resistor. This time, set it for the most appropriate range for reading the resistance. Your goal is to use the lowest scale that will work for the resistor. If you read from the bands that you have a 2.2kΩ resistor, the lowest scale that will work is the 20k scale. If you set it on the next one down, the 2k scale, the resistance will be too large to be read on that range.
- If you have a piece of gear available to work on, open it up and look for the resistors. Last time you did this, you just measured the resistors without identifying their values. Now you have the tools to read their band patterns and determine their values. You can then use your DMM to confirm that your reading of the band pattern is correct. Don't be concerned if a few of them show apparent values that are significantly lower than expected, as this can happen in a circuit. There may be other components that also provide apparent resistance in parallel. As we will see later, this can substantially change the overall resistance.

This exercise has given you experience with the resistor color code, both with identifying unknown resistors from their band patterns and by determining what colors to look for when a particular value is desired.

Answers

Resistor Color Codes

Fourth example: 15.8kΩ, 1% (5-band) = brown–green–gray–red–brown

Fifth example: 2.2MΩ, 5% (4-band) = red–red–green–gold

Sixth example: 6.49kΩ, 1% (5-band) = blue–yellow–white–brown–brown

Seventh example: 750kΩ, 5% (4-band) = violet–green–yellow–gold

Eighth example: 57.6Ω, 1% (5-band) = green–violet–blue–gold–brown

Unknown Resistor Identification

Resistor #1: gray–red–gold–gold = 8.2Ω±5%

Resistor #2: red-white-yellow-brown-brown = 2.94kΩ±1%

Resistor #3: brown–orange–orange–gold = 13kΩ±5%

Resistor #4: yellow–gray–violet–orange–brown = 487kΩ±1%

Resistor #5: orange–blue–green–gold = 3.6MΩ±5%

Resistor #6: orange–yellow–black–yellow–brown = 3.40MΩ±1%

8

Power and Watt's Law

Goal: When you have completed this chapter, you will have a working knowledge of how the property of power is related to voltage, resistance, and current.

Objectives

· Define power and its unit.
· Relate power to voltage and current with Watt's law.
· Use Watt's law to calculate the power being used in a simple circuit.
· Combine Watt's law and Ohm's law to relate power directly to voltage and resistance.
· Demonstrate power being used in a circuit.

Power

So far we have become familiar with the three properties of a circuit that are governed by Ohm's law: voltage, current, and resistance. Voltage is defined as how much potential energy a group of electrons possesses and will release if they are allowed to return to the ground. Current is defined as the number of electrons that are passing a certain point on their way to the ground. Resistance is defined as the amount of restriction the electrons encounter on their way to the ground. We will now add a fourth property to this list. This property is called *power*. Together, these four properties are the main attributes of electronic circuits.

Power is defined as a measurement of how much work is being done by electrons when they encounter a resistance. It is symbolized by a capital **P**. The unit of measurement of power is the *watt*, abbreviated **W**. It is named for James Watt (1736–1819), who contributed substantially to the development of the steam engine after it was invented by Robert Fulton. He also observed the relationship between potential, current, and power, so Watt's law is named after him. In physics, power is defined as the rate at which work is done by any system, not just electronics. We'll confine ourselves to the discussion of power in the context of electronics.

Power in a Circuit

Power is used by a circuit when it converts the electrical energy of the movement of electrons into some other kind of energy. There are a number of types of energy to choose from. The most common are heat, light, acoustic energy, and various types of electromagnetic energy. Most of the time, we are concerned with the waste heat that is generated by electronic components.

· Heat is generated by most electronic components, particularly resistors and transistors.

· Light is generated by LEDs and lamps.

- Acoustic energy is generated by loudspeakers and piezo elements.
- A magnetic field is generated by a coil, such as a transformer or inductor.
- Radio waves are generated by specific configurations of electronic components.

Measuring and Calculating Power with Watt's Law

While there are direct ways of measuring the power being used by a particular system, these methods can be cumbersome and inaccurate (unless they are performed in a lab under rigorous controls). It's always more convenient to calculate how much power is being used by a system than to measure it. We do this by finding out the quantities that are easier to measure: voltage, current, and resistance. If we look at a circuit with a certain voltage across it (E) and a certain current flowing through it (I), we can then calculate the power being used by the circuit, abbreviated P.

Here's the equation for Watt's law, which relates power to current and voltage:

P = I × E (power = current × voltage)

Just as with Ohm's law, the units of the quantities you use in the power equation must be correct for the result to be in the correct units: potential (voltage) must be in volts (V), current must be in amps (A), power must be in watts (W). Therefore we have to remove any prefixes before we reach for our calculators. Also, just as with Ohm's law, we can use a formula circle for Watt's law, as shown in figure 8-1.

Figure 8-1.

The formula circle is a quick and easy way to find the three rearrangement equations of the Watt's law equation: P = I × E, I = P / E, and E = P / I. To find a Watt's law equation using the circle, just use your finger or thumb to cover up the quantity that you are calculating, and the relationship of the other two quantities will be revealed. For example, if we want to calculate the current from a known power and voltage, we cover the "I" of the circle. The two remaining letters are the P and the E, with the P on top of the E. From this, we obtain the equation I = P / E, which is one of the three arrangements of Watt's law listed above.

Just as an example of how this is useful, let's look at the power being used by the first circuit you built in the first hands-on practice. This circuit is shown again in figure 8-2.

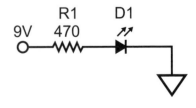

Figure 8-2.

We would like to know the power (P) being used by the circuit. Watt's law is P = I × E, so we need to know the current (I) and the voltage (E). In the later hands-on practice with the DMM, you measured both of these. The current was found to be 16mA, and the voltage was found to be 9.0V. (Your actual numbers were probably a little different, but we'll use these for the example.) We'll use four steps for the calculations, as shown in Appendix B.

Quantities: We list the quantities that we know, and the one quantity that we're going to solve for. If there are any prefixes, we remove them. Here we need to remove the "milli" prefix from the quantity of 16mA.

I = 16mA = 0.016A
E = 9.0V
P = ???

Equation: Identify the equation we need, and rearrange this equation so we can solve for the quantity that we don't know. Here we know the current (I) and the voltage (E), and we said above that we want to calculate the power (P). Watt's law is the equation that relates power, current, and voltage. We do not need to rearrange, because we already have all the known quantities on one side and the single unknown on the other.

P = I × E

Solve: Combine the quantities that you know with the equation, and solve the math.

P = 0.016A × 9.0V
P = 0.144W

Prefix: Add a prefix if one is needed. 0.144 is smaller than 1, so we need a prefix.

P = 144mW

When you built the circuit, the total power being used was just a little more than 1/8 of a watt.

Making Watt's Law Easier to Use

If we know the current of a circuit, we can put it into the Watt's law equation (as we just did) to calculate the power. Current is not nearly as easy to measure directly, so it's much more likely that we will only know the voltage and the resistance.

We could simply use these two attributes to calculate the current (using Ohm's law) and plug that value into the Watt's law equation for power. This is a perfectly valid method of solving the question but it takes two sets of calculations.

We could also do some substitution of Watt's law and Ohm's law to construct an equation for power that we can plug our values for voltage and resistance into.

First, we rearrange Ohm's law to solve for current (I):

E = I × R

I = E / R

Then we substitute E / R for I in Watt's Law:

P = I × E

P = (E / R) × E

We could use this equation just as it is, or we can rearrange it to get the voltage terms together:

$$P = (E \times E) / R$$

$$P = E^2 / R$$

We can show how this works by just looking at the power being used by the resistor in the first circuit. Some of the power in the circuit is being turned into light (and some into heat) by the LED. The rest is being turned into heat by the resistor. In the DMM exercise, you measured the actual voltage across the resistor, and you measured the actual resistance. Here we'll demonstrate the combined Watt's law and Ohm's law equations on the nominal values of these two quantities, a voltage of 7.5V across the resistor and a resistance of 470Ω. We'll use four steps, as shown in Appendix B.

Quantities: We list the quantities that we know, and the one quantity that we're going to solve for. If there are any prefixes, we remove them.

E = 7.5V
R = 470Ω
P = ???

Equation: Identify the equation we need, and rearrange this equation so we can solve for the quantity that we don't know. Here we know the voltage (E) and the resistance (R), and we'll use our rearranged and substituted Watt's law equation from above. We do not need to rearrange, because we already have all the known quantities on one side and the single unknown on the other:

$$P = E^2 / R$$

Solve: Combine the quantities that you know with the equation, and solve the math.

P = (7.5V)² / 470Ω
P = 56.25V² / 470Ω
P = 0.1196W

Prefix: Add a prefix if one is needed. 0.1196 is smaller than 1, so we need a prefix. We can round a bit if we want to, to make the number more reasonable to talk about.

P = 119.6mW
P = 120mW

This is about 1/8 of a watt. Looking at this calculation and at the earlier one for the entire circuit, we see that 144mW is being used by the circuit, of which 120mW is being dissipated as heat by the resistor. Only 24mW is being used by the LED.

Applications for Watt's law

There are a number of situations where it's important to know the amount of power being used by a particular circuit. Often there are safety concerns: all of the components in the circuit need to be rated to handle the amount of current being run through them. In the calculations we just did, we saw that the 470Ω resistor in the first hands-on example circuit is dissipating 1/8W. This size of resistor is rated at 1/4W, so it's still within a safe range. A cautious designer would probably use a 1/2W resistor for this application if it was on a soldered circuit board.

If we do the calculations on the power use of a 100Ω resistor in the same position, we get a very different result. The voltage drop will stay essentially the same in this new circuit, so the calculations are as follows.

Quantities: We list the quantities that we know, and the one quantity that we're going to solve for. If there are any prefixes, we remove them.

E = 7.5V
R = 100Ω
I = ???

Equation: Identify the equation we need, and rearrange this equation so we can solve for the quantity that we don't know. Here we know the voltage (E) and the resistance (R), and we'll use our rearranged and substituted Watt's law equation from above. We do not need to rearrange, because we already have all the known quantities on one side and the single unknown on the other:

$P = E^2 / R$

Solve: Combine the quantities that you know with the equation, and solve the math.

$P = (7.5V)^2 / 470Ω$
$P = 56.25V^2 / 470Ω$
$P = 0.5625W$

Prefix: Add a prefix if one is needed. 0.5625 is smaller than 1, so we need a prefix. We can round a bit if we want to, to make the number more reasonable to talk about.

P = 562.5mW
P = 560mW

This is a little over 1/2W, and we just said that the resistor is rated at 1/4W. Building the circuit with a 100Ω resistor wouldn't result in an explosion or anything so dramatic, but if you try it (as my students occasionally do, by accident) you will find that the resistor becomes quite warm to the touch and can even burn you.

Hands-On Practice

This exercise will give you experience calculating power with Watt's law.

Materials

These are the materials you will need for these exercises.

- Digital multimeter (DMM)
- Breadboard
- 9V battery and battery clip with wires
- 100Ω resistor
- Red LED
- Insulated solid wire (20, 22, or 24 gauge)

- We just did the calculations to show how much power is being dissipated by the 100Ω resistor in one of our standard test circuits. We got a result of 560mW, just over 1/2W.

- The circuit is shown in figure 8-3.

- Use the resistor color code to identify a 100Ω resistor.

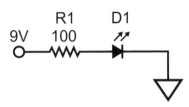

- Build the circuit as shown.

- Use your DMM to measure the voltage of the supply voltage bus to the ground. This should be about 9V.

- Measure the voltage across the LED. This should be about 1.5V.

Figure 8-3.

- Measure the voltage across the resistor. This should be about 7.5V.

- Write down this reading.

- Change your DMM to measure current on the 20mA scale.

- Open the circuit and measure the current. This should read about 16mA.

- Change your DMM to read resistance on the 200Ω scale.

- Leaving the circuit open, measure the resistance across R1. This should be about 100Ω.

- Reconnect the circuit so that the LED is lit.

- Feel the resistor by holding it gently between your finger and thumb. You should feel the resistor start to heat up after a minute or two. Let go if it gets too hot to touch.

- As shown in the calculations earlier, this heat dissipation is approximately 1/2W, and these resistors are rated at 1/4W. The amount of heat you are feeling is about twice what the resistor can handle on a long-term basis. In the short term, it just gets a lot warmer than it should, and it will burn up and fail if we leave it there.

- Clear the breadboard.

Ohm's Law and Watt's Law Calculations

Set 1

Use Ohm's law to calculate the current in the circuit shown in figure 8-4. Use the four steps, as shown in the chapter. The answer is given at the end of the exercise.

Now that you know the current, use Watt's law to calculate the power being used in the same circuit. Use the four steps, as shown in the chapter. The answer is given at the end of the exercise.

Figure 8-4.

Set 2

Use the rearrangement of Watt's law with voltage and resistance (no current) to calculate the power being used in the circuit shown in figure 8-5. Use the four steps, as shown in the chapter. The answer is given at the end of the exercise.

Use Ohm's law to calculate the current in the circuit shown. Use the four steps, as shown in the chapter. The answer is given at the end of the exercise.

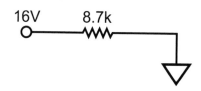

Figure 8-5.

Set 3

Use Watt's law to calculate the power being used in the circuit shown in figure 8-6. Use the four steps, as shown in the chapter. The answer is given at the end of the exercise.

Use Ohm's law to calculate the resistance in the same circuit. Use the four steps, as shown in the chapter. The answer is given at the end of the exercise.

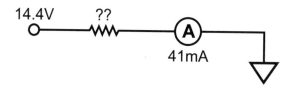

Figure 8-6.

Set 4

Use Ohm's law to calculate the resistance in the circuit shown in figure 8-7. Use the four steps, as shown in the chapter. The answer is given at the end of the exercise.

Use Watt's law to calculate the power being used in the same circuit. The answer is given at the end of the exercise.

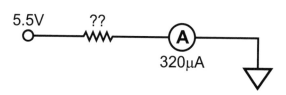

Figure 8-7.

Set 5

Use Ohm's law to calculate the voltage in the circuit shown in figure 8-8. Use the four steps, as shown in the chapter. The answer is given at the end of the exercise.

Now that you know the voltage, use Watt's law to calculate the power being used in the same circuit. Use the four steps, as shown in the chapter. The answer is given at the end of the exercise.

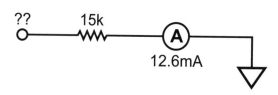

Figure 8-8.

You have built a circuit that demonstrates how much power a resistor dissipates, and you have gotten experience calculating power with Watt's law, while also reviewing Ohm's law calculations.

Answers

Ohm's Law and Watt's Law Calculations

Set 1

Quantities
   E = 14V
   R = 4.7kΩ = 4700Ω
   I = ???
Equation
   I = E / R
Solve
   I = 14V / 4700Ω
   I = 0.0029787A
Prefix
   I = 3.0mA

Quantities
   I = 3.0mA = 0.003A
   E = 14V
   R = ???
Equation
   P = I × E
Solve
   P = 0.003A × 14V
   P = 0.042W
Prefix
   P = 42mW

Set 2

Quantities
   E = 16V
   R = 8.7kΩ = 8700Ω
   P = ???
Equation
   $P = E^2 / R$
Solve
   $P = (16V)^2 / 8700Ω$
   $P = 256V^2 / 8700Ω$
   P = 0.029425W
Prefix
   P = 30mW

Quantities
   E = 16V
   R = 8.7kΩ = 8700Ω
   I = ???
Equation
   I = E / R
Solve
   I= 16V / 8700Ω
   I = 0.001839A
Prefix
   I = 1.8mA

Set 3

Quantities
   E = 14.4V
   I = 41mA = 0.041A
   P = ???
Equation
   P = I × E
Solve
   P = 0.041A × 14.4V
   P = 0.5904W
Prefix
   I = 590mW

Quantities
   E = 14.4V
   I = 41mA = 0.041A
   R = ???
Equation
   R = E / I
Solve
   R = 14.4V / 0.041A
   R = 351.22Ω
Prefix
   R = 351Ω

Set 4

Quantities
    E = 5.5V
    I = 320μA = 0.00032A
    R = ???
Equation
    R = E / I
Solve
    R = 5.5V / 0.00032A
    R = 17,187.5Ω
Prefix
    R = 17kΩ

Quantities
    E = 5.5V
    I = 320μA = 0.00032A
    P = ???
Equation
    P = I × E
Solve
    P = 0.00032A × 5.5V
    P = 0.00176W
Prefix
    I = 1.76mW

Set 5

Quantities
    I = 12.6mA = 0.0126A
    R = 15kΩ = 15,000Ω
    E = ???
Equation
    E = I × R
Solve
    E = 0.0126A × 15,000Ω
    E = 189V
Prefix
    E = 189V

Quantities
    I = 12.6mA = 0.0126A
    E = 189V
    P = ???
Equation
    P = I × E
Solve
    P = 0.0126A × 189V
    P = 2.3814W
Prefix
    P = 2.4W

9

Total Resistance in Series and Parallel

Goal: When you have completed this chapter, you will have an understanding of how resistances interact when they are combined in series and parallel circuits.

Objectives

- Use subscripts correctly in an equation.
- Calculate the total resistance of resistors in series.
- Calculate the total resistance of resistors in parallel.
- Measure the total resistance of resistors in series in a circuit.
- Measure the total resistance of resistors in parallel in a circuit.
- Observe the change in current in a circuit when a resistor is added in parallel.

The Larger View of Resistors in Series and Parallel

Knowing how resistors act when we combine them in series and in parallel is an important concept in electronics. This interaction is also the key to understanding other concepts that we will learn later, such as the interaction of impedances and the function of a potentiometer to control voltage levels. This chapter introduces more math to help demonstrate the interactions between resistors, but the math is far less important than the underlying concepts.

Using Subscripts

In this chapter we discuss total resistance, for which we use the term R_t. Here the lowercase "t" is written below the line and in smaller type. This is called a *subscript*. We use a subscript to modify a symbol for a term and make it more specific. For example, in the equation $E = I \times R$, the R stands for resistance, any resistance. If you see R_t in an equation, you know that it doesn't just stand for just any resistance; it stands for total resistance, with the "t" standing for *total*. In the voltage divider equation, we are looking at the relationship between the input voltage and the output voltage. We need to be able to tell them apart in the equation, so we use E_{in} for input voltage and E_{out} for output voltage. By adding the subscript, we've made the term more specific: not just *voltage*, but *input voltage*.

We'll also use this when we have several resistors or other components acting like resistors. In an equation, we designate three resistances as R_1, R_2, and R_3. This designation is only used in the equations where we are working with more than one resistance. In a circuit diagram, the resistors are still labeled **R1, R2,** and **R3,** just as we have been doing all along.

Resistors

Resistors, with their function of limiting current, are one of the most common components found in electronic circuits. When we discuss resistors and how they interact with each other, we need to bear in mind that the discussion extends to objects that act equivalently to resistors, such as the resistance of a piece of cable. The rules given here for resistors in parallel and series also apply to these resistances.

Resistors may be used alone, in combination, and in conjunction with other components to perform a wide variety of functions. Understanding the basics of how they function when they are combined is one of the keys of understanding how the circuit performs overall. Resistors can be combined in two ways, in series and parallel. All other groupings of resistors can be analyzed and found to be combinations of these two connections. (If you take an electrical engineering course, you'll find that EE folks are fascinated with analyzing weird and complex configurations of resistors that are never found in real-life circuits. I have yet to find a practical use for this kind of analysis, even though I love puzzles.)

The total resistance is designated R_t when we see it in an equation. We need to be able to mentally simplify what's going on in a circuit, or just to be able to view a complex circuit as a simple circuit. For example, let's say we want to buy a power supply (a "wall-wart" or battery eliminator) for a guitar pedal. We need to know what voltage and current the power supply needs to provide. Finding the voltage is easy. You can look at the jack where the power supply gets plugged in, and the voltage will usually be listed there. You can look on the spec sheet or the label on the underside to find out how much current it needs. If this is not listed, you can use an ammeter to measure how much current the pedal allows through when it's doing the work we want it to do (making a guitar sound like a UFO, for example). This is called the *current draw*, the amount of current flowing from the supply to the ground.

We can think of a complex circuit as acting like a single resistance. If we open up the pedal or look at its circuit diagram, we'll see a confusing mess of stuff: resistors, capacitors, diodes, LEDs, transistors, and op amps, and maybe a voltage regulator, an inductor, or an optoisolator. All of this stuff is connected together in a much more complicated circuit than we could hope to analyze at this point. Fortunately, for now we don't really need to figure out how it works. We just need to know how much it limits the current when the circuit components are all used together. We can think of the pedal as a single resistor that is letting through a certain amount of current when presented with a certain voltage supply.

When we go to the electronics store, we need to look for a voltage supply that has the ability to provide at least as much current as our pedal wants. If it can provide more, that's fine. I once had a salesperson at a well-known electronics chain try to convince me that the 9V adapter I was buying was too powerful and would damage my electronics by forcing too much current through. It was rated at the correct voltage, with a much higher current rating. Here is where a good working knowledge of Ohm's law is important: if the voltage supply is the same, and the resistance is staying the same, there's no reason that the current will be different. The voltage is the pressure pushing the current through. If the voltage of the adapter is the same as that of the battery, no more current will flow through with the pedal.

We can see that we can think of a complex circuit as being a simple resistance, as we said earlier. This concept is part of *Thévenin's theorem*, which says that any set of voltage sources and resistances can be thought of as a single voltage source and a single resistance. This leads us to a discussion of total resistance, where we look at several resistors in series or in parallel and calculate what single resistor could be substituted for them.

Resistors in Series

Here's the quick rule for resistors in series: When resistors are in series, the total resistance R_t is always more than the largest individual resistance.

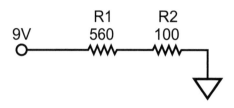

In the circuit shown in figure 9-1, the total resistance of the circuit is the resistance of the two resistors added together. We can think of resistors as a little slalom course for electrons, with traffic cones laid out that they have to weave around to get through. The 100Ω resistor has 100 cones, and the 560Ω resistor has 560 cones. If the electrons have to go through both of the resistors one after the other, they'll encounter all 660 of the cones, so they'll be slowed down even more than by either of them separately. Whether they are in two groups or a single one, they will still slow down the traffic the same amount.

Figure 9-1.

When R_t is the total resistance, and R_1 and R_2 are the individual resistors in a series connection, that gives us this equation:

$$R_t = R_1 + R_2$$

If we added another resistor to this, such as R3, then they would all add together to give this equation:

$$R_t = R_1 + R_2 + R_3$$

We can make a general equation for any number (n) of resistors:

$$R_t = R_1 + R_2 + \ldots + R_n$$

In other words, we say that the total resistance of resistors in series is the sum of their individual resistances. Note that we use the words "in series" because the situation is different when the resistors are in parallel. When you calculate R_t, there is a quick rule to check your answer: When resistors are in series, the total resistance R_t is always more than the largest individual resistor.

Calculating R_t for Series Resistors

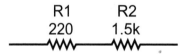

Here is an example problem: Find the total resistance R_t of two resistors in series, as shown in figure 9-2.

As usual, we'll use four steps for the calculations, as shown in Appendix B.

Figure 9-2.

Quantities: We list the quantities that we know, and the one quantity that we're going to solve for. If there are any prefixes, we remove them.

$R_1 = 220\Omega$
$R_2 = 1.5k\Omega = 1,500\Omega$
$R_t = ???$

Equation: Identify the equation we need, and rearrange this equation so we can solve for the quantity that we don't know.

$$R_t = R_1 + R_2$$

Solve: Combine the quantities that you know with the equation, and solve the math.

$$R_t = 220\Omega + 1{,}500\Omega$$
$$R_t = 1{,}720\Omega$$

Prefix: Add a prefix if one is needed.

$$R_t = 1.72\text{k}\Omega$$

This is a long way of just adding two resistor values together. It's shown formally for the sake of consistency, to get us ready for the more complex calculations ahead, and also to remind you that the equation only works if the prefixes are removed before starting.

An Audio Application: Speakers in Series

A speaker is different from a resistor because its resistance changes depending on the frequency it is being asked to reproduce. Consequently we measure a speaker's impedance rather than its resistance. Impedance (Z) is also measured in ohms (Ω), and the same rules are used for calculating the total impedance as for total resistance: the total impedance (Z_t) of impedances in series is equal to the sum of all the impedances ($Z_1, Z_2,$ and so on.).

If we have a single 4Ω speaker hooked up to an amplifier set at a certain level, we'll hear a certain level of signal. If we hook up two 4Ω speakers in series, the total impedance becomes $4\Omega + 4\Omega = 8\Omega$. Ohm's law works for audio signals, and since we've increased the impedance (resistance), the current must decrease, less work will be done, and the signal level will drop. This is the opposite of what we would expect: if we add a second speaker, we expect our signal to be louder, but it gets quieter instead.

Resistors in Parallel

Here's the quick rule for resistors in parallel: When resistors are in parallel, the total resistance R_t is always less than the smallest individual resistance. It may seem counterintuitive that we can get a smaller resistance from two resistors, but that's what happens. Let's look at how this works.

We start with a simple circuit of a resistor R1 in series with a battery, as shown in figure 9-3. A certain amount of current will flow through the resistor. Ohm's law lets us calculate how much: $I = E / R = 9V / 560\Omega = 0.016A = 16\text{mA}$.

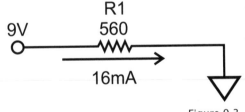

Figure 9-3.

If we add a second resistor of the same value, R2, so that the resistors R1 and R2 are arranged in parallel, the current has another path available to it. Each of these paths has the same resistance, so each individually has the same current as the original circuit with just one resistor. By adding a second resistor, we've offered another path, and the current takes advantage of it. Now we have two resistors with 16mA flowing through each one. This is shown in figure 9-4.

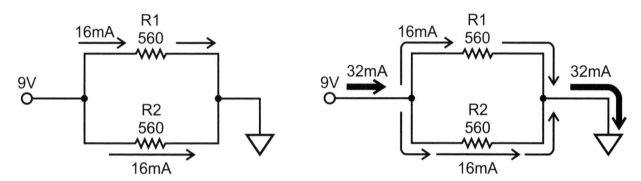

Figure 9-4.   Figure 9-5.

Now we have an application of Kirchhoff's current law. Each resistor has 16mA flowing through it. Kirchhoff's current law says that the current entering a point must equal the current leaving that point. The point where the resistors are connected to the (+) side of the battery must have 16mA flowing out of it toward R1 and an additional 16mA flowing out of it toward R2. Therefore there must be 32mA flowing into that point. This is shown in figure 9-5. When the two arms of the circuit come back together, the 16mA in each arm must add up to 32mA.

Now we go back to Ohm's law. We still have a 9V battery, and we have 32mA flowing out of it. Knowing these two things, we can calculate how much resistance the battery is seeing in the circuit: R = E / I = 9V / 0.032A = 280Ω. It's as if there is just a single 280Ω resistor instead of the two resistors that are actually there. From the two 560Ω resistors in parallel, we got a total resistance that is lower than either one. This is shown in figure 9-6. The battery doesn't care if there are two 560Ω resistors or a single 280Ω resistor, because they both look the same to the battery.

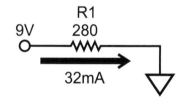

Figure 9-6.

Another way to look at this is to imagine the water tank again. If we have a huge tank full of water, and we open the valve of a pipe at the bottom of it, we'll get a certain flow of water. Let's imagine a flow of 10 gallons per second. If we then open the valve on a second pipe of the same size, we'll get an additional 10 gallons per second. With both taps open, we get 20 gallons per second. A person who can't see the valves but just sees how much water is flowing past has no idea whether there's one big pipe or two smaller ones or even more whose flow adds up to 20 gallons per second. It's the same with resistors: for each one we plug into the voltage source, the more current flows out.

From this observation, we can conclude the quick rule of thumb given earlier for checking your answer when you calculate the total resistance: when resistors are in parallel, the total resistance R_t is always less than the smallest individual resistance.

The Total Resistance of Resistors in Parallel

For the total resistance of two resistors in parallel, there are two equations. These are the same mathematically, but they look very different. First there is a short-form equation for just two resistors in parallel:

$$R_t = \frac{(R_1 \times R_2)}{(R_1 \times R_2)}$$

Here's the long-form equation for three or more resistors in parallel:

$$\frac{1}{R_t} = \frac{1}{R_1} + \frac{1}{R_2} + \frac{1}{R_3} + \cdots + \frac{1}{R_n}$$

This equation can be rearranged to give R_t on one side:

$$R_t = \frac{1}{\dfrac{1}{R_1} + \dfrac{1}{R_2} + \dfrac{1}{R_3} + \cdots + \dfrac{1}{R_n}}$$

Any of these forms are acceptable, but the short form is easier to use. Note that if you have more than two resistors in parallel, the short-form equation does not work and you must use the long one.

Calculating R_t for Parallel Resistors

Here is an example problem, to find the total resistance R_t of the resistors in the circuit shown in figure 9-7.

We'll do this with both forms of the equation, just to prove to ourselves that we really do get the same answer each time and that both equations work. First we'll use the two-resistor form of the equation:

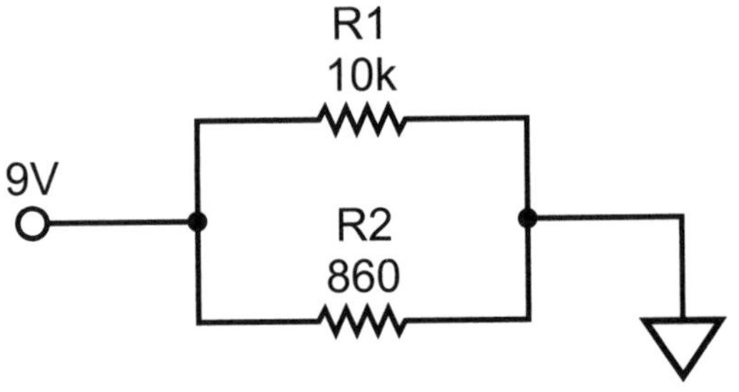

Figure 9-7.

Quantities: We list the quantities that we know, and the one quantity that we're going to solve for. If there are any prefixes, we remove them.

$$R_1 = 10k\Omega = 10{,}000\Omega$$
$$R_2 = 860\Omega$$
$$R_t = ???$$

Equation: Identify the equation we need, and rearrange this equation so we can solve for the quantity that we don't know.

$$R_t = (R_1 \times R_2) / (R_1 + R_2)$$

Solve: Combine the quantities that you know with the equation, and solve the math.

$$R_t = (10{,}000\Omega \times 860\Omega) / (10{,}000\Omega + 860\Omega)$$
$$R_t = (8{,}600{,}000) / (10{,}860)$$
$$R_t = 791.8968\Omega$$

Prefix: Add a prefix if one is needed. Here we don't need a prefix, because the answer is already between 1 and 999.

$R_t = 792\Omega$

This answer meets the criteria of our rule of thumb that the total resistance R_t will be smaller than either of the resistances that make it up. Now we'll try the same calculation with the general form of the equation, which works for any number of resistors:

Quantities: We list the quantities that we know, and the one quantity that we're going to solve for. If there are any prefixes, we remove them.

$R_1 = 10k\Omega = 10,000\Omega$
$R_2 = 860\Omega$
$R_t = ???$

Equation: Identify the equation we need, and rearrange this equation so we can solve for the quantity that we don't know. Here we rearrange to put R_t on the left.

$(1 / R_t) = (1 / R_1) + (1 / R_2)$
$R_t = 1/((1 / R_1) + (1 / R_2))$

Solve: Combine the quantities that you know with the equation, and solve the math.

$R_t = 1/((1 / 10,000\ \Omega) + (1 / 860\Omega))$
$R_t = 1/((0.0001) + (0.0011627906))$
$R_t = (1 / 0.0012627906)$
$R_t = 791.8968\Omega$

Prefix: Add a prefix if one is needed. Here we don't need a prefix, because the answer is already between 1 and 999.

$R_t = 792\Omega$

An Audio Application: Speakers in Parallel

As we saw earlier in the audio example, impedances act like resistances for calculating the total impedance (Z_t) of impedances in parallel. Also, just as before, we'll get a certain signal level from a certain amplifier with a single 4Ω speaker.

If we hook up two 4Ω speakers in parallel, the total impedance becomes 2Ω. We've decreased the impedance by a factor of two, so the current increases by a factor of two. Watt's law tells us that power (P) is equal to current (I) times voltage (E). If we double the current, we double the power. This gives us a higher signal level from the speakers. The danger is that we will stress the amplifier by doing this. If it's a 100W amp and we were only using 75W when we had just one speaker, with both speakers the amp would be trying its best to put out 150W. This is a recipe for amplifier disaster.

We need to be aware of this because it's all too easy to hook up speakers in parallel by accident. Passive floor monitors for live sound and some PA speakers have two jacks in the back for 1/4" TS plugs. If you plug the TS connector from the power amp into one and a TS cable from another speaker to the other, now you have the speakers connected in parallel. If that's what you intended and the power amp can handle it, then all is well. If not, the amplifier will shut down or potentially even melt down.

A Brief Summary

When we connect resistors in series and in parallel, we get completely opposite results in terms of total resistance. It's important to keep this in mind as we move forward, because every time we build or analyze a circuit we'll be looking at resistors or components acting like resistors that are interacting according to these two rules. We don't need to do the calculations every time, but we do need to remember the essence of how resistors interact.

- R_t for resistors in series is greater than the largest of the individual resistors.
- R_t for resistors in parallel is less than the smallest of the individual resistors.

Combining Series and Parallel

Not all circuits are so neatly laid out that there are only series connections or only parallel connections. More often the resistors (or other components) are arranged in a combination of series and parallel. The way to deal with this is to solve for the parts that we can visualize as being just in series or just in parallel. Consider the circuit in figure 9-8, where we would like to know how much current will be flowing. To calculate the current, we need to know the total resistance. This is an example of Thévenin's theorem, which was mentioned earlier: we don't need to calculate individual currents in each arm of the circuit or anything like that; we can just figure out how to look at this mass of resistors as a single resistor.

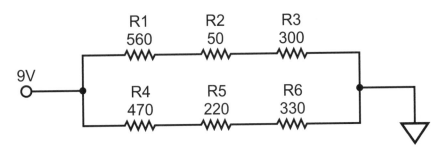

Figure 9-8.

This example is a combination of series and parallel: R_1, R_2, and R_3 are in series (R_{123}); R_4, R_5, and R_6 are in series (R_{456}); these two resistances are in parallel. If we wanted to find the total resistance R_t for the circuit, we would solve for R_{123} and then solve for R_{456}. Finally we would find R_t by solving for R_{123} and R_{456} in parallel.

$R_{123} = 560\Omega + 50\Omega + 300\Omega$
$R_{123} = 910\Omega$

$R_{456} = 470\Omega + 220\Omega + 330\Omega$
$R_{456} = 1020\Omega$

$R_t = (R_{123} \times R_{456}) / (R_{123} + R_{456})$
$R_t = (910\Omega \times 1020\Omega) / (910\Omega + 1020\Omega)$
$R_t = (928200) / (1930)$
$R_t = 481\Omega$

Any complex network of resistances can be analyzed in terms of series and parallel until we arrive at a single total resistance. We will find this idea very useful when we are looking at how transistors work, at how impedances interact, how voltage dividers function, and many other concepts. While calculating total resistance is a matter of using the math, the math is just the tool we use to discover the specifics of what we already know: that resistances in series act as a larger resistance, and resistances in parallel act as a smaller resistance.

Hands-On Practice

This exercise helps you gain experience building parallel circuits and measuring their resistance and current.

Materials

These are the materials you will need for these exercises.

- Digital multimeter (DMM)
- Breadboard
- 9V battery and battery clip with wires
- 470Ω resistor
- 560Ω resistor
- Two 1.0kΩ resistors
- Red LED
- Insulated solid wire (20, 22, or 24 gauge)

Resistors in Series

- Use the resistor color code to identify a 470Ω resistor and a 560Ω resistor.

- Before starting circuit construction, use your DMM to measure the exact resistance of these two resistors R1 and R2. Write down these values on a sheet of paper.

- Calculate the total resistance R_t of these two resistors in series. Use the four steps (Quantities, Equation, Solve, Prefix) for this calculation. (Yes, this is a lot of unnecessary writing just to add two numbers together, but it's good practice for using the steps later for using the equation for resistors in parallel.)

- Check your answer against the sample calculation that is shown at the very end of the exercise. While your answer will probably not be exactly the same, it should be pretty close to the sample calculation.

- Set up your breadboard with a 9V battery.

- Construct the circuit shown in figure 9-9 on your breadboard. The LED should light.

- Disconnect the circuit by unplugging the LED from the ground.

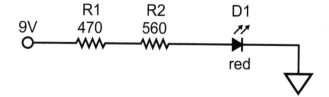

Figure 9-9.

- The circuit for placing the meter to measure total resistance R_t between nodes A and B is shown in figure 9-10. The circuit diagram shows that the circuit is broken between the LED and the ground.

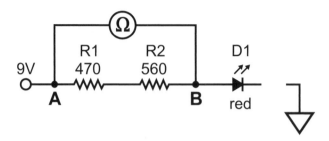

Figure 9-10.

- The measurement technique is shown in figure 9-11.

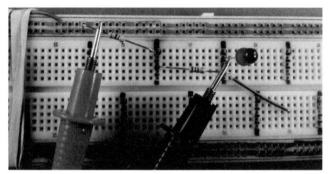

Figure 9-11.

- Compare the total resistance you measured with the total resistance you calculated earlier. They should be the same, within a few ohms. If they are not, check that you have built the circuit properly and measured it in the right place.
- Clear the breadboard for building the next circuit.

Resistors in Parallel, Circuit 1

- Use the resistor color code to identify a 560Ω resistor and a 1.0kΩ resistor.
- Before starting circuit construction, use your DMM to measure the exact resistance of the two resistors R1 and R2. Write down these values on a sheet of paper.
- Calculate the total resistance R_t of these two resistors in parallel. Use the four steps (Quantities, Equation, Solve, Prefix) for this calculation.
- Check your answer against the sample calculation that is shown at the very end of the exercise. While your answer will probably not be exactly the same, it should be pretty close to the sample calculation.

- Construct the circuit shown in figure 9-12 on your breadboard. The LED should light.

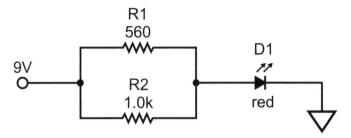

Figure 9-12.

- Prove to yourself that the resistors in the circuit are in parallel. You can do this by unplugging one end of one resistor. If the resistors are in parallel, as they should be, the light from the LED should dim a little but not go out entirely.

- Reconnect the resistor to restore the circuit to parallel operation.

- Disconnect the circuit by unplugging the LED from the ground.

- The DMM connections for measuring the total resistance of these two resistors in parallel is shown in figure 9-13. Again, notice that the circuit is broken.

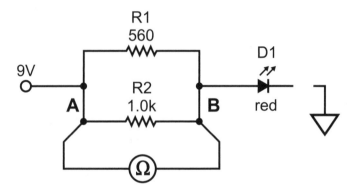

Figure 9-13.

- Measure the total resistance R_t between nodes A and B as shown in figure 9-14.

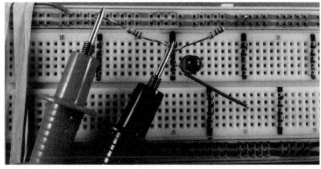

Figure 9-14.

- Compare the total resistance you measured with the total resistance you calculated earlier. If these are not the same within a few ohms, check that you have built the circuit properly. If you need a reminder of how to build the parallel circuit, you can go back to the earlier hands-on exercise.

- Clear the breadboard for building the next circuit.

Resistors in Parallel, Circuit 2

- Use the resistor color code to identify a 470Ω resistor and two 1.0kΩ resistors.

- Before starting circuit construction, use your DMM to measure the exact resistance of the resistors R1, R2, and R3. Write down these values on a sheet of paper.

- Use these resistance values to calculate the total resistance R_t of resistors R1 and R2 in parallel. Use the four steps (Quantities, Equation, Solve, Prefix) for this calculation.

- Check your answer against the sample calculation that is shown at the very end of the exercise. While your answer will probably not be exactly the same, it should be pretty close to the sample calculation.

- Construct the circuit shown in figure 9-15 on your breadboard.

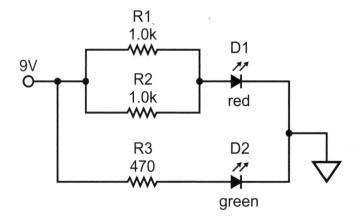

Figure 9-15.

- Prove to yourself that the resistors R1 and R2 in the circuit are in parallel. You can do this by unplugging one end of one of them. If the resistors are in parallel, as they should be, the light from the red LED D1 should dim a little but not go out entirely. The green LED D2 should not be affected.

- Reconnect the resistor to restore the circuit.

- Prove to yourself that the two arms of the circuit are in parallel. You can do this by unplugging one end of the red LED (D1), plugging it back in, and then unplugging one end of the green LED (D2). If the arms of the circuit are in parallel, as they should be, the LED connected with that arm should go out, and the other should stay on.

- Reconnect the LED to restore the circuit.

- Disconnect the parallel circuit by unplugging one end of the red LED (D1) from the ground.

- Measure the total resistance R_t between nodes A and B as shown in figure 9-16.

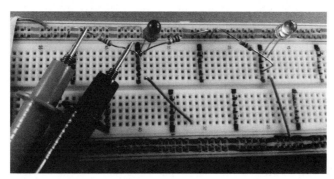

Figure 9-16.

- Compare the total resistance you measured with the total resistance you calculated earlier. If these are not the same within a few ohms, check that you have built the circuit properly. If you need a reminder of how to build the parallel circuit, you can go back to the earlier hands-on exercise.
- Restore the circuit to full function with both LEDs lit.

Reviewing Ohm's Law

- You have just shown yourself that the resistances of R1 and R2 together in parallel are very close to the resistance of R3 by itself.
- Ohm's law says that current is the voltage divided by the resistance. Since the resistance in each arm is very similar, we expect that that the currents through each arm of the circuit should also be very close.
- Calculate the current in the first arm that consists of the parallel arm of R1, R2, and D1. Sample calculations are shown at the end.
- Calculate the current in the second arm that consists of R3 and D2. Again, sample calculations are shown at the end.
- With the calculations complete, you will measure the current in the second arm of the circuit, consisting of R3 and D2.
- Disconnect (unplug) the wire connecting the short lead of green LED D2 from the ground bus.
- Plug this free wire into an unused component strip of the breadboard.
- Set your DMM to measure current on the 20mA scale.

Figure 9-17.

- Measure the current between the free wire of D2 and the ground. This is shown in figure 9-17. The green D2 LED should light up while the meter is connected.

- Write down this current with its units. This is the current in the D2 arm of the circuit.

- Reconnect the short lead of D2 to the ground.

- Now measure the current in the first arm of the circuit, consisting of R1, R2, and D1.

- Disconnect (unplug) the wire connecting the short lead of LED D1 from the ground bus.

- Plug this free wire into an unused component strip of the breadboard.

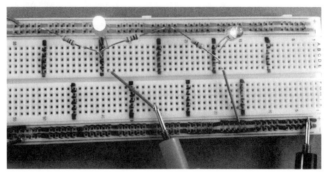

Figure 9-18.

- Measure the current between the free lead of D1 and the ground. This is shown in figure 9-18. The red D1 LED should light up while the meter is connected.

- Write down this current with its units. This is the current in the D1 arm of the circuit.

- As discussed earlier, we expect that the two current measurements will be the same, or very similar in value. Is this the case? Both should be as shown in the sample calculations at the end of the exercise.

- We can see a further example of Ohm's law by unplugging one end of the resistor R1. When this is done on a complete circuit, the LED dims. If you do it on this circuit with the LED disconnected for current measurement, and then measure the current, the current shown on the meter should be half of what you see with both resistors in place. This is shown in figure 9-19.

Figure 9-19.

- Clear the breadboard.

This set of hands-on exercises has given you some experience with calculating and measuring resistances in series and parallel. It's unusual to find resistors in parallel and series like these simple circuits, but the concept of how resistances in parallel and series interact with each other is one of the central concepts of electronics.

Answers

Resistors in Series

Quantity

$R_1 = 470\Omega$

$R_2 = 560\Omega$

$R_t = ???$

Equation

$R_t = R_1 + R_2$

Solve

$R_t = 470\Omega + 560\Omega$

$R_t = 1030\Omega$

Prefix

$R_t = 1.03k\Omega$

Resistors in Parallel, Circuit 1

Quantity

$R_1 = 560\Omega$

$R_2 = 1.0k\Omega = 1000\Omega$

$R_t = ???$

Equation

$R_t = (R_1 \times R_2) / (R_1 + R_2)$

Solve

$R_t = (560\Omega \times 1000\Omega) / (560\Omega + 1000\Omega)$

$R_t = (560,000) / (1560)$

$R_t = 358.97\Omega$

Prefix

$R_t = 360\Omega$ (no prefix needed)

Resistors in Parallel, Circuit 2

Quantity

$R_1 = 1.0k\Omega = 1000\Omega$

$R_2 = 1.0k\Omega = 1000\Omega$

$R_t = ???$

Equation

$R_t = (R_1 \times R_2) / (R_1 + R_2)$

Solve

$R_t = (1000\Omega \times 1000\Omega) / (1000\Omega + 1000\Omega)$

$R_t = (1,000,000) / (2000)$

$R_t = 500\Omega$

Prefix

$R_t = 500\Omega$ (no prefix needed)

Ohm's Law, First Arm of R1, R2, and D1

Quantity

$E = 9V$

$R = 500\Omega$

$I = ???$

Equation

$I = E / R$

Solve

$I = 9V / 500\Omega$

$I = 0.018A$

Prefix

$I = 18mA$

Ohm's Law, Second Arm of R3 and D2

Quantity

$E = 9V$

$R = 470\Omega$

$I = ???$

Equation

$I = E / R$

Solve

$I = 9V / 470\Omega$

$I = 0.01914A$

Prefix

$I = 19mA$

Voltage Dividers, Linear and Audio Potentiometers

10

Goal: When you have completed this chapter, you will be familiar with potentiometers and how they act as voltage dividers to reduce signal levels.

Objectives

· Build a circuit with resistors configured as a voltage divider.
· Recognize how input and output impedances act together as a voltage divider.
· Understand the function of a potentiometer as a variable-resistance voltage divider.
· Recognize the symbol for a potentiometer.
· Measure the difference between a linear taper potentiometer and an audio taper potentiometer.
· Test an unknown potentiometer to determine its value and taper.
· Observe the variable voltage output from a potentiometer used in a circuit.

The Voltage Divider

The *voltage divider* is significant because it is the first application of components we've looked at where we take two or more components and arrange them in a particular way to perform a certain task. The concept of the voltage divider is essential to understanding the function of potentiometers, filters, and impedance.

A voltage divider is made up of two resistances in series. The function of a voltage divider is to take an input voltage and reduce it by a certain proportion. When we look at a circuit diagram, we may be able to discover two resistors in series and know that they are being used as a voltage divider to provide a reduced voltage to a particular point in the circuit. It's one of the easiest "building block" configurations of components to recognize. We'll start by studying the circuit shown in figure 10-1.

When current is running through a single resistor, the full voltage can be measured at the input end, as shown in the diagram. The voltage measurement shown will be 9V. If we measure on the other side of the resistor, the voltage will be zero. This is shown in figure 10-2.

So what can we say about the voltage in the middle of the resistor? We could try to cut a little hole in the resistor to measure the voltage there, but that would ruin the resistor. Fortunately, there is an easier way: since any resistance is the sum of all the resistances in series, we could replace

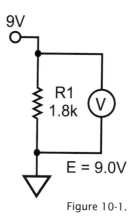

Figure 10-1.

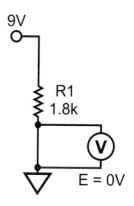

Figure 10-2.

the 1.8kΩ resistor with two 900Ω resistors. The total resistance R_t is the same, but now we can see what's going on inside the 1.8kΩ resistor, as shown in figure 10-3.

If we have a potential of 9V at one end of the 1.8kΩ resistor and 0V at the other end, in the middle we would see half of 9V, which is 4.5V.

It follows that anywhere we choose to look in a resistor (or series of resistors), the voltage will be proportional to the resistance that remains between the point where we are and the ground. This sounds a lot more complicated than it really is. It just means that if the current has passed a quarter of the way through the resistor, the voltage is down by a quarter. Three-quarters of the total resistance remains in front of it, so at that point the voltage will be three-quarters of what we started with. Figure 10-4 shows a graphic representation of the voltage level inside a resistor that's hooked up to a 9V battery at one end and ground at the other. Signal ground is defined as being 0V.

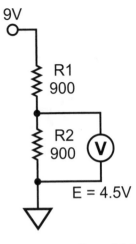

Figure 10-3.

The Voltage Divider Equation

The *voltage divider* is made up of two resistors, R1 and R2. R1 is closer to the input, and R2 is tied to the signal ground. Together in series, these make up a total resistance of R_t. We'll recall from the earlier discussion of resistances in series that the total resistance R_t is simply these two resistances added together: $R_t = R_1 + R_2$.

For any input voltage Ein, the output voltage E_{out} will be proportional to however much R_2 contributes to the total resistance R_t. For example, if R_2 is one-eighth of the total resistance, then the output voltage will be one-eighth of the input voltage.

This relationship between the voltages and resistances gives us the following equation:

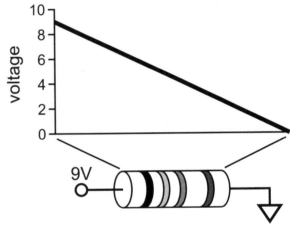

Figure 10-4.

$$E_{out} = E_{in} \times \left(\frac{R_2}{R_1 + R_2} \right)$$

Let's see this equation in action on the voltage divider in figure 10-5. As always, we will use the four-step method to solve for the answer:

Quantities: We list the quantities that we know, and the one quantity that we're going to solve for. If there are any prefixes, we remove them.

$E_{in} = 12V$
$R_1 = 300\Omega$
$R_2 = 150\Omega$
$E_{out} = ???$

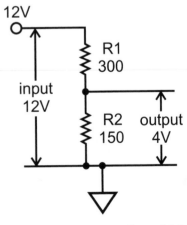

Figure 10-5.

Equation: Identify the equation we need, and rearrange this equation so we can solve for the quantity that we don't know.

$$E_{out} = E_{in} \times (R_2 / (R_1 + R_2))$$

Solve: Combine the quantities that you know with the equation, and solve the math.

$E_{out} = 12V \times (150\Omega / (300\Omega + 150\Omega))$
$E_{out} = 12V \times (150\Omega / (450\Omega))$
$E_{out} = 12V \times (0.3333)$
$E_{out} = 4V$

Prefix: Add a prefix if one is needed. Here we don't need a prefix, because the answer is already between 1 and 999.

$E_{out} = 4V$

What we see happening here is that the potential, the voltage level, between the two resistors is proportional to the resistance relative to the total resistance. R2 is 150Ω, which is one-third of the total resistance, so the output voltage across that one resistor is one-third of the total input voltage. It's pretty easy to prove to ourselves that whatever voltage we plug into this voltage divider, we'll get one-third of that voltage as the output voltage.

We've just looked at the output of a particular voltage divider. Now we'll look at the output of a voltage divider where the resistors are both ten times larger than the one we just looked at. This is shown in figure 10-6.

Will the output be ten times larger, or ten times smaller, or just the same? Here's the math:

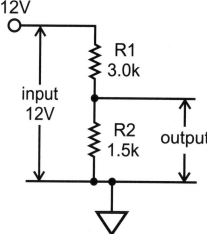

Figure 10-6

Quantities: We list the quantities that we know, and the one quantity that we're going to solve for. If there are any prefixes, we remove them.

$E_{in} = 12V$
$R_1 = 3.0k\Omega = 3000\Omega$
$R_2 = 1.5k\Omega = 1500\Omega$
$E_{out} = ???$

Equation: Identify the equation we need, and rearrange this equation so we can solve for the quantity that we don't know.

$$E_{out} = E_{in} \times (R_2 / (R_1 + R_2))$$

Solve: Combine the quantities that you know with the equation, and solve the math.

$E_{out} = 12V \times (1500\Omega / (3000\Omega + 1500\Omega))$
$E_{out} = 12V \times (1500\Omega / (4500\Omega))$
$E_{out} = 12V \times (0.3333)$
$E_{out} = 4V$

Prefix: Add a prefix if one is needed. Here we don't need a prefix, because the answer is already between 1 and 999.

$E_{out} = 4V$

Even though the resistors are ten times larger, the output voltage E_{out} stays exactly the same. This is because the output is dependent on the proportion between the resistors, not the actual values of the resistors. Any two resistors where R_2 is half the resistance of R_1 will give the same result.

The Limitations of a Voltage Divider

While a voltage divider is a very useful configuration for reducing voltage, it does have some limitations. First, the output voltage is a certain percentage of the input, so the input voltage needs to be stable for the output voltage to remain stable. If the input voltage dips or spikes, so does the output.

A more serious limitation of the voltage divider has to do with what we use the output voltage for. If our goal is simply to have a particular voltage at a particular point and not to actually use any of that potential, there is no problem at all. Usually, we want to use the voltage provided by the divider to run something or do something. Let's say we want to use the 4V output of the voltage divider that we just created above to supply some current to charge up a cell phone. If we plug our phone into a 4V charger, we can measure the current and find that the phone is drawing 25mA.

Remember that we can think of the phone as just being a simple resistor: at a certain voltage (4V), it is allowing through a certain amount of current (25mA). Ohm's law can tell us what resistance the phone is acting like.

Quantities: We list the quantities that we know, and the one quantity that we're going to solve for. If there are any prefixes, we remove them.

E = 4V
I = 25mA = 0.025A
R = ???

Equation: Identify the equation we need, and rearrange this equation so we can solve for the quantity that we don't know.

E = I × R
R = E / I

Solve: Combine the quantities that you know with the equation, and solve the math.

R = 4V / 0.025A
R = 160Ω

Prefix: Add a prefix if one is needed. Here we don't need a prefix, because the answer is already between 1 and 999.

R = 160Ω

When we plug our cell phone into the 4V supply, the load of the cell phone will look like a 160Ω resistor to the voltage divider. This is shown in figure 10-7. Now R_2 isn't the only resistance across the output; we also have the load resistance R_{load}. R_2 and R_{load} are in parallel, and together they are providing a lower resistance than R_2 was by itself. We can calculate the total resistance R_t of these two resistances.

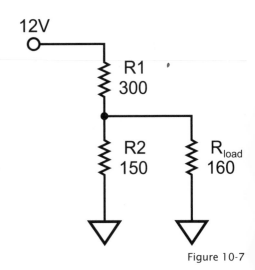

Figure 10-7

Quantities: We list the quantities that we know, and the one quantity that we're going to solve for. If there are any prefixes, we remove them.

R_{load} = 160Ω
R_2 = 150Ω
R_t = ???

Equation: Identify the equation we need, and rearrange this equation so we can solve for the quantity that we don't know. This is the standard R_t equation for two resistors, with R_{load} substituted for R1.

$R_t = (R_{load} \times R_2) / (R_{load} + R_2)$

Solve: Combine the quantities that you know with the equation, and solve the math.

$R_t = (160\Omega \times 150\Omega) / (160\Omega + 150\Omega)$
$R_t = (24,000) / (310)$
$R_t = 77\Omega$

Prefix: Add a prefix if one is needed. Here we don't need a prefix, because the answer is already between 1 and 999.

$R_t = 77\Omega$

This is going to change everything in the voltage divider, because R_2 isn't 150Ω anymore; now it's 77Ω. We can do the calculations to find out what the output voltage will be with this new situation.

Quantities: We list the quantities that we know, and the one quantity that we're going to solve for. If there are any prefixes, we remove them.

E_{in} = 12V
R_1 = 300Ω
R_2 = 77Ω
E_{out} = ???

Equation: Identify the equation we need, and rearrange this equation so we can solve for the quantity that we don't know.

$E_{out} = E_{in} \times (R_2 / (R_1 + R_2))$

Solve: Combine the quantities that you know with the equation, and solve the math.

$E_{out} = 12V \times (77\Omega / (300\Omega + 77\Omega))$
$E_{out} = 12V \times (77\Omega / (377\Omega))$
$E_{out} = 12V \times (0.2042)$
$E_{out} = 2.45V$

Prefix: Add a prefix if one is needed. Here we don't need a prefix, because the answer is already between 1 and 999.

$E_{out} = 2.45V$

This shows that as useful as the voltage divider is for providing a certain voltage from a higher one, we must also be very careful not to load up that voltage source too much, or the voltage itself will drop. It's not that the 12V supply isn't keeping up its end of the bargain, but rather that we've dropped the output voltage by putting a large load across it. We started with a 4V supply, and we ended up with a 2.5V supply. A voltage divider is an inefficient way of providing a particular voltage, because its output changes substantially if a load of any size is placed on it. We will see this demonstrated in the hands-on exercise. Fortunately there are other ways of providing a consistent voltage to charge a cell phone or do other desired work for us, and voltage dividers have other functions.

The Voltage Divider as a Model

We've just seen that the real-world voltage divider can be very useful for reducing a voltage proportionally but that it has some limitations: in order to stay close to the expected output, the load must remain small. The voltage divider also has a purpose when we are thinking about how electronic components interact with each other. Many electronics situations can become much easier to think about if we regard them as being in the form of a voltage divider. For example, in a moment we will look at potentiometers, which we can regard as being variable voltage dividers. Later in the chapter, we will look at the interaction of output and input impedances, which become much clearer when we put them together and think of them as a voltage divider. We can also use the voltage divider as a model for understanding electronic filters composed of capacitors, inductors, and resistors. The behavior of transistors and tubes and op amps can all be examined this way also. The voltage divider is useful as a real-world configuration of two resistors, but its greater value is in how we use it to understand the field of electronics as a whole.

The Audio Pad

The voltage divider can also be used with AC signals. We saw earlier that the voltage divider reduces the voltage to a particular portion of the original. It also reduces voltage, regardless of whether it's positive or negative. From moment to moment, the voltage divider reduces the AC voltage by whatever proportion the voltage divide reduces a DC voltage. A voltage divider used on an audio signal is also called a *pad*. An example of this is shown in figure 10-8.

This is often depicted in the configuration shown in figure 10-9.

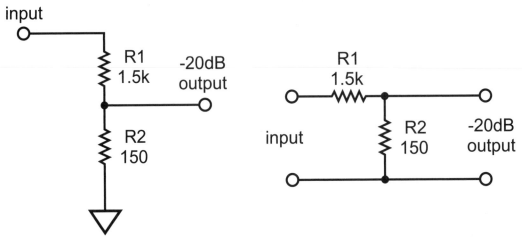

Figure 10-8.

Figure 10-9.

Impedance

We defined *impedance* (Z) as the ability to restrict current flow when it changes with frequency. Among the challenging aspects of understanding audio electronics are the concepts of input and output impedance and how they interact. We can use the voltage divider as a way of understanding how an output impedance of an audio source works with the input impedance of an audio amplifier.

Output Impedance

Every audio source (microphone, preamp, console out, power amplifier out) has the ability to restrict the current that flows out of it. We'll use the example of a dynamic microphone, but the theory works the same for all audio sources. As shown in figure 10-10, when sound waves hit the microphone capsule, the coil moves in and out and generates a voltage across the two wires coming off the capsule. If the capsule isn't connected to anything, no current can move anywhere. The voltage is still there.

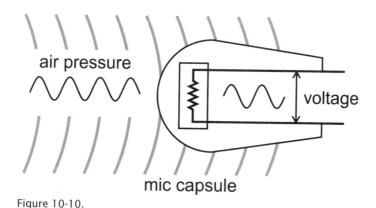

Figure 10-10.

If we connect the two wires from the capsule to an external impedance, now the current can go somewhere. How much current moves off the capsule depends on two things: how much internal impedance the capsule has and how much external impedance the capsule is hooked up to. Every audio source has an internal impedance, which is how much it restricts the current that comes off of it. We call this the *output impedance* because it's the impedance associated with the output of the source. In the figure, this has been shown as a resistor inside the capsule. This impedance is an intrinsic property of the capsule, based on a number of physical and electrical factors. Because it is impedance, it acts differently over a range of frequencies. At one particular frequency, it acts equivalently to a particular resistor. Ohm's law governs how much current can flow out of the source, based on the resistance and the voltage of the source at that frequency.

Input Impedance

The other parts of the impedance situation are the devices that a source can be connected to, such as mic preamps, line-level inputs, loudspeakers, and so on. Just as the sources have the ability to restrict current, so do the inputs that we connect them to. This ability to restrict the current flowing into a device is called the *input impedance*. While the actuality of the input circuitry is much more complex, we can visualize this impedance as simply being a single resistor between the input and the signal ground. This is shown in figure 10-11. Ohm's law governs how much current flows into the device, based on the voltage presented to the input and the impedance at that frequency.

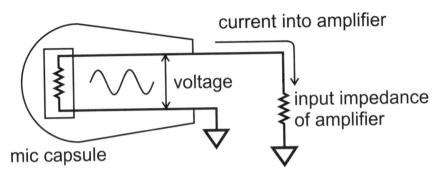

Figure 10-11.

Impedances Interacting as a Voltage Divider

When an audio source (such as a microphone) is connected to an amplification stage (such as a mic preamplifier), their impedances act as a voltage divider. As shown in figure 10-12, the source's output impedance Z_{out} acts like R_1 in a voltage divider. Similarly, the amplification stage's input impedance Z_{in} acts like R_2 in a voltage divider.

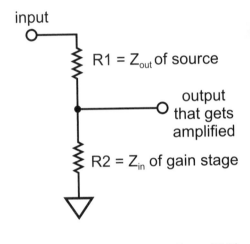

Figure 10-12.

If the input impedance Z_{in} is lower than the output impedance Z_{out} of the source that is connected to it, whatever voltage the source is generating will be substantially reduced. From this interaction, we obtain the rule of thumb that the input impedance of an input stage should always be at least 10 times higher than the impedance of the output stage.

If you've ever tried plugging an electric guitar with a Z_{out} of 20kΩ into a line-level input of 20kΩ, you'll have experienced the consequences of impedance mismatch: a thin-sounding, very low-level signal. A quick glance at the impedances reveals that only about a half of the voltage from the guitar pickup will be available for amplification. The rest apparently gets lost, or we can think of it as being stuck in the instrument. It's not that the voltage has actually disappeared; it's just that it's become a victim of the input impedance being too low.

Potentiometers

Variable resistors are usually called *potentiometers*, or *pots*. Like other components, they are made in a dizzying array of sizes, shapes, and specifications. Several features will usually distinguish a pot from other components.

• It has 3 contacts.

• It has some method of adjusting the position of the center contact (called the wiper) somewhere along the resistive material between the outer two contacts; usually this is a rotating shaft, but sometimes it's a slider, such as those on a mixing board.

• It's marked to indicate how much resistance it has.

The old and new symbols for a pot are shown in figure 10-13.

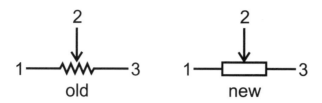

Figure 10-13.

The resistive material is between contacts 1 and 3; it is between these two points that the listed value of the pot is measured. Contact 2 is the wiper. Some pots are shown in figure 10-14. The pots shown in the figure are designed to be soldered onto a circuit board, while others have solder lugs for connecting to wires. On a circuit diagram, a pot may be labeled **R** (for resistor), **VR** (for variable resistor), or **P** (for potentiometer).

Figure 10-14.

Pots as Voltage Dividers

We can use a pot as a variable voltage divider. The balance between the two resistors of the voltage divider can be adjusted by the wiper. The part of the resistive element between the top (pin 1) and the wiper (pin 2) is like R_1. The part of the resistive element between the wiper (pin 2) and the bottom (pin 3) is like R_2. This is shown in figure 10-15.

We can still think of the pot as being two resistances, R_1 and R_2. When the pot is turned up, R_1 becomes small, and R_2 becomes large, approaching the value of R_t. The output is the ratio R_2 / R_t, and this ratio becomes essentially 1, so all the available input is going to the output.

When the pot is turned down, R_1 becomes large and R_2 becomes small. As R_2 gets smaller and approaches zero, the output level falls and approaches zero.

In many pots, the resistive material between the contacts is consistent, so that the amount of resistance between one end and the wiper has a linear relationship: changing the position by half of a turn changes the resistance by one half. For this reason, these are called *linear taper* pots.

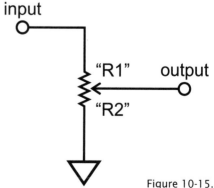

Figure 10-15.

If we use a linear pot on an audio signal, a strange thing happens. Let's look at what happens with a 4V signal. With the pot set halfway down, the two resistances are the same: $R_1 = R_2$. We don't need to know the actual resistances, because they are equal. With this information, we can calculate the output voltage at this setting.

Quantities: We list the quantities that we know, and the one quantity that we're going to solve for. If there are any prefixes, we remove them.

$E_{in} = 4V$
$R_1 = R_2$
$E_{out} = ???$

Equation: Identify the equation we need, and rearrange this equation so we can solve for the quantity that we don't know.

$E_{out} = E_{in} \times (R_2 / (R_1 + R_2))$

Solve: Combine the quantities that you know with the equation, and solve the math.

$E_{out} = 4V \times (R_2 / (R_2 + R_2))$
$E_{out} = 4V \times (R_2 / 2 \times R_2)$
$E_{out} = 4V \times (1/2)$
$E_{out} = 2V$

Prefix: Add a prefix if one is needed. Here we don't need a prefix, because the answer is already between 1 and 999.

$E_{out} = 2V$

As we would expect, when the pot is set at the halfway point, the output voltage is half of the input voltage. With this information, we can now calculate the gain using the gain equation, which we haven't seen before. The equation gives us the *voltage gain* (A_e) from the input voltage (E_{in}) and the output voltage (E_{out}). This gain equation gives us an answer in decibels (dB). Here, "A" stands for amplitude, and the "e" subscript stands for voltage.

$$A_e = 20 \times \log \left(\frac{E_{out}}{E_{in}} \right)$$

Quantities: We list the quantities that we know, and the one quantity that we're going to solve for. If there are any prefixes, we remove them.

$E_{in} = 4V$
$E_{out} = 2V$
$A_e = ???$

Equation: Identify the equation we need, and rearrange this equation so we can solve for the quantity that we don't know.

$A_e = 20 \times \log (E_{out} / E_{in})$

Solve: Combine the quantities that you know with the equation, and solve the math.

$A_e = 20 \times \log (2V / 4V)$
$A_e = 20 \times \log (0.5)$
$A_e = 20 \times (-0.301)$
$A_e = -6.02 \ dB$

Prefix: Add a prefix if one is needed. Here we don't need a prefix, because the answer is already between 1 and 999.

$$E_{out} = -6dB$$

What this answer shows us is that when the pot is set to half and the output voltage is half, the level is 6dB down from full. If we set the pot to a quarter, the level goes down by half again, and when we do the math again, we find the level is another 6dB down. As shown in figure 10-16, this makes the pot very insensitive at the top end and very sensitive at the bottom end. On the console, we're used to faders that sound equally sensitive along their entire length. These are called *audio taper* pots, and they are used extensively in audio equipment.

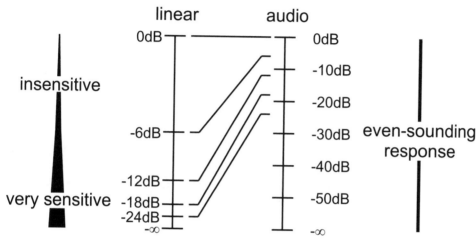

Figure 10-16.

The linear pot has consistent resistance over its length. The audio pot has variable resistance over its length that gives it a logarithmic response. The resistive material is narrow at one end and wider at the other. This corresponds to the way we hear: we perceive a sound that has twice the amplitude as being only 6dB louder. Similarly we perceive a sound that has half the amplitude as being 6dB quieter.

To sum up, the point of using an audio taper pot is that it sounds right when we are adjusting sound levels, even though the resulting voltage levels look very uneven when we watch them on the voltmeter or oscilloscope. If we use a linear taper pot for controlling audio signal levels, it will sound wrong.

Identifying the Value and Taper of an Unknown Pot

Most pots are labeled with their value and their taper, **B** for linear and **A** for audio. Sometimes we need to identify a pot that is unlabeled (or is labeled in a way that is so arcane as to be indecipherable). Two important questions can be asked about an unknown pot: what resistance it has (its value) and whether it is an audio pot or a linear pot. We can answer both of these with an ohmmeter.

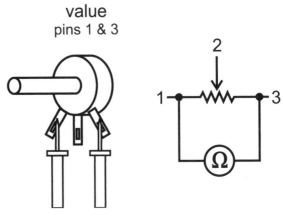

value
pins 1 & 3

Figure 10-17.

We can find the resistance (or value) of the pot by measuring the resistance between pin 1 and pin 3, the two resistive pins. This is shown in figure 10-17. No matter where we set the wiper, this value will be the same.

The resistance curves of audio and linear pots are shown in figure 10-18. These graphs show how the resistance between pins 1 and 2 varies as the shaft is turned or the slider is moved. The linear pot has a straight line, as the name implies. The audio pot has a logarithmic resistance. If we look at the resistance at the 50% point on each graph, this will show us how to address measuring the taper of an unknown pot. When a linear pot is turned up to the halfway point, it has 50% of its total resistance. When an audio pot is turned up to the halfway point, it only has about 10% of its total resistance.

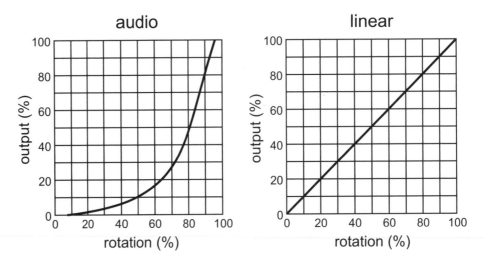

Figure 10-18.

This difference in resistance at the halfway point means that we can determine the taper of a pot by setting the pot at its center point and measuring the resistance from the wiper to each end. If the values are the same or similar (50-50 or somewhere close to that, such as 40-60), then the pot is linear. If the values differ by a factor of 10 or so, then the pot is audio. This procedure is shown in figure 10-19. The hardest part of this is getting the pot set to the center point. This is usually a matter of turning the shaft back and forth several times to determine the clockwise and counterclockwise limits of movement. Most pots have a slot or flat on the shaft that will help with this. Having found how far around the shaft can turn, move the shaft to the midpoint

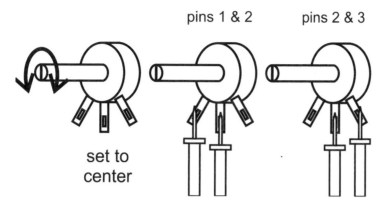

pins 1 & 2   pins 2 & 3

set to
center

Figure 10-19.

between the two limits. Don't worry if this point is a little bit off. Most pots are not manufactured to especially accurate standards, so the exact center of rotation is unlikely to be the exact center of the resistance. This is why a 40-60 split is still considered to be a linear pot.

How Pots Are Used as Level Controls

In most audio gear, the electronics that perform the amplification have a set amount of gain. When we need to control the amount of signal from that amplification stage, we use a potentiometer as a variable voltage divider. An audio taper pot can be used to reduce the level of signal being amplified, controlling the input level. An audio taper pot can also be used to restrict the level of the output signal from the gain stage. These are typical controls on microphone preamps, for example. Not every preamp has both of these controls. These two controls are shown in figure 10-20. The amplification stage is shown by a simple triangle. Another example is the trim knob on the console channel strip controlling the input level, while the fader controls the output level.

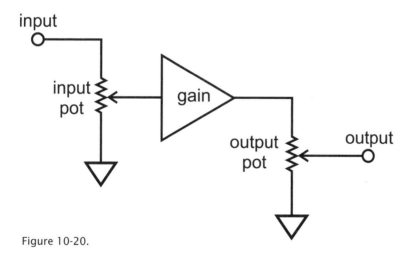

Figure 10-20.

The input pot is used to control the signal level to keep it from overdriving the gain stage. The output pot is used to control the level of the amplified signal. Pots have many other functions in audio electronics, but their most obvious and prominent use is to control signal levels.

Hands-On Practice

The goal of this exercise is to become familiar with fixed voltage dividers and potentiometers. You will learn to identify a pot's value and taper using a DMM. You will also examine the difference between an audio taper pot and a linear taper pot by measuring their resistance at several positions through their rotation.

Materials

These are the materials you will need for these exercises.

- Digital multimeter (DMM)
- Breadboard
- 9V battery and battery clip with wires
- 470Ω resistor
- Two 1.0kΩ resistors
- 22kΩ resistor
- 10kΩ linear taper pot
- 10kΩ audio taper pot
- Four jumper cables with alligator clips
- Solid insulated wire (20, 22, or 24 gauge) about 12" long
- Optional: grab bag of potentiometers

Voltage Divider

- Use the resistor color code to identify a 1.0kΩ resistor and a 22kΩ resistor.
- Before starting circuit construction, use your DMM to measure the exact resistance of the two resistors R1 and R2. Write down these values on a sheet of paper.
- Set up your breadboard with a 9V battery.
- Construct the voltage divider circuit shown in figure 10-21 on your breadboard.

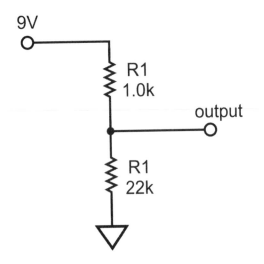

Figure 10-21.

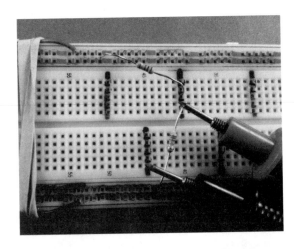

Figure 10-22.

- Measure the supply voltage (from the voltage supply bus to the ground). You should get a reading of about **9V**. Write down this reading on the sheet of paper.
- Use this voltage reading and the two resistance readings to calculate the voltage output from the voltage divider. Use the four steps (Quantities, Equation, Solve, Prefix) for this calculation.
- Check your answer against the sample calculation that is shown at the very end of the exercise. While your answer will probably not be exactly the same, it should be pretty close to the sample calculation.
- Measure the voltage of the voltage divider output, as shown in figure 10-22. This is the voltage across R2.
- Compare this voltage with the voltage divider output you calculated earlier. The two should be identical, within a few tenths of a volt. If they are not, check that you have built the circuit properly and measured it in the right place.
- Leave the circuit set up for the next exercise.

Loaded Voltage Divider

- Use the resistor color code to identify a 1.0kΩ resistor.
- Construct the circuit shown in figure 10-23 on the breadboard. The LED should light up, but it will likely be fairly dim.
- Measure the voltage of the voltage divider output, as shown in figure 10-24. This is the voltage across R2. You should get a reading of around 5V or 5.5V.
- This circuit demonstrates what happens when a voltage divider output is loaded. A voltage divider is only really useful at maintaining a particular voltage if little or no load is placed on it.
- Clear the breadboard for building the next circuit.

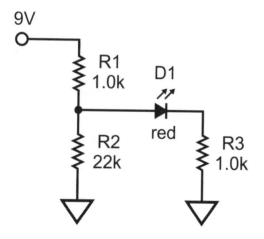

Figure 10-23.

Figure 10-24.

Unknown Potentiometer Identification

You already know the resistance and taper of your two pots, but we're going to imagine for a moment that these pots are unlabeled and need identification.

- Choose one of the two pots at random.

- Use your DMM to measure the resistance between the two outer terminals. This is shown in figure 10-25. It can be a little tricky to hold the pot, the red probe, and the black probe, and to then get the probes to make good contact with the terminals while avoiding touching the probes with your fingers.

Figure 10-25.

- Write down the reading you get on a piece of paper, including the unit and the prefix if you need one. This is the resistance of the pot, also called its value. You expect to see a value of about 10kΩ. Inexpensive pots often have substantial differences from their listed values, such that it would not be surprising to get a reading of 12kΩ or 7.4kΩ for this "10kΩ" pot.

- The shaft on a typical pot has either a flat spot on its side or a slot down the center, as shown in figure 10-26. If your pot doesn't have either of these, for this next step it is helpful to put an identifying mark on the shaft so you know what position the shaft is in. A little flag made of tape works well for this.

Figure 10-26.

- Turn the shaft or knob of the pot back and forth several times to get a feeling for how far it turns in each direction. This is shown in figure 10-27.

Figure 10-27.

- Based on these observations, set the pot in the center, as shown in figure 10-28. For the pot shown, this position is with the slot horizontal.

Figure 10-28.

- Measure the resistance between the left outer terminal and the wiper terminal, in the center, as shown in figure 10-29. Write down the reading with its value and prefix.

Figure 10-29.

- Measure the resistance between the wiper terminal and the right outer terminal, as shown in figure 10-30. Write down the reading with its value and prefix.

- If these two values are pretty close to the same, the pot has a linear taper. If they are quite different, the pot has an audio taper. Write down your answer on your paper.

- If you have found the linear pot, set it aside for the next part of the exercise.

- Use the same procedure to measure the taper of the other pot: measure the total resistance at the two outer pins, set the shaft in the center, and measure and compare the resistance from the center to the side and the center to the other side. You expect to find a pot of the same value but with the other taper, since you should have one audio taper pot and one linear taper pot.

Figure 10-30.

Comparing Linear and Audio Pots

The object of this section is to demonstrate how a real linear pot and a real audio pot behave when we examine their resistance at several points in the range of motion. You will be collecting 14 points of data and using them to draw a small graph. You can collect the data in a table and draw it on regular paper following the format of the data table and blank graph shown below, or you can print out the blank data table and blank graph from the supplemental materials on the DVD.

- Choose the linear taper pot that you identified in the previous part of the exercise.

- Turn the shaft as far clockwise as it will go, as shown in figure 10-31. This is position 1.

Figure 10-31.

- Set the DMM to read resistance on the 20kΩ scale.
- Use a jumper cable with alligator clips to attach one DMM probe to pin 1 (on the left), as shown in figure 10-32. Make sure that the alligator clip is only making contact with pin 1 and is not touching pin 2 (the center contact) or the metal housing of the pot. Since you will be measuring resistance, it doesn't matter which probe is used for this connection.

Figure 10-32.

- Use a second jumper cable with alligator clips to attach the other DMM probe to pin 2 (in the center), as shown in figure 10-33. Check that there are no accidental connections.

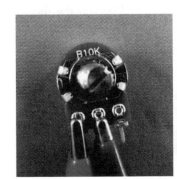

Figure 10-33.

- The blank data table is shown here in table 10-1.

shaft position	linear taper resistance (kW)	audio taper resistance (kW)
1 (up full)		
2		
3		
4 (center)		
5		
6		
7 (down full)		

Table 10-1.

- You should get a resistance reading equal to the resistance value of the pot that you determined earlier, 10kΩ or similar. Write the value in the first open space in the data table, position 1 for the linear taper pot. This data is all expressed in kΩ, so you just need to write down the number from the DMM display.
- This is the first data point for the linear pot. It shows the resistance of the pot when it is turned up full. When the pot is used as a voltage divider for controlling an audio signal, this setting would give no attenuation of the signal. In terms of the voltage divider, R1 is zero and R2 is the total resistance, so the output is as high as it can be.

- Turn the pot's shaft about one-eighth of a turn to position 2, as shown in figure 10-34. For most pots, this is the position with the flat facing to the right, or with the slot vertical.

- Read the resistance at this position and record it in your data table as position 2 for the linear pot.

- Continue turning the shaft one-eighth of a turn to each position, followed by reading the resistance and recording the reading in your data table for each position.

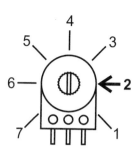

Figure 10-34.

- Position 7 is the last, with the shaft turned all the way counterclockwise, as shown in figure 10-35.

- The resistance at this position should give a zero reading on the DMM. Record this reading in the chart on the last line for the linear pot.

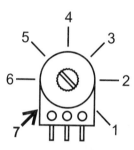

Figure 10-35.

- Plot these data points for the resistances on the blank graph, as shown in figure 10-36. Use a solid dot for each of the data points.

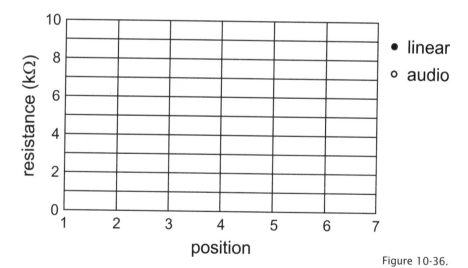

Figure 10-36.

- Connect the dots with straight lines. Since this is a linear pot, the graph of the resistance will approximate a straight line. (The chances of it being a completely straight line are quite small, so don't expect it to be.)

- Repeat the procedure with the audio taper pot: set it full clockwise, attach it to the DMM, and take the readings at the seven points.

- Plot these data points on the blank graph also, using an open circle for each point. Connect the dots. What you see in this graph is logarithmic resistance that will give a smooth audio signal response.

Linear Pot in a Circuit

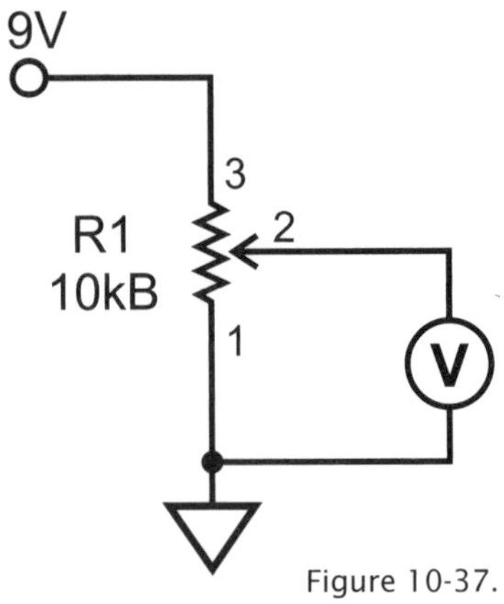

In this exercise you will examine how a pot can be used as part of a circuit.

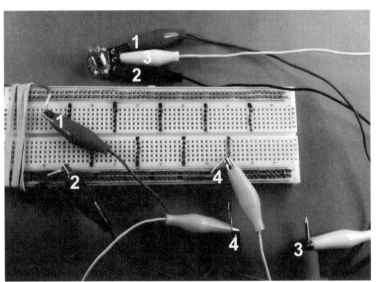

Figure 10-37.

- The circuit diagram for this test setup is shown in figure 10-37. Note that the pot is labeled **R1–10kB**: here "10k" stands for 10kΩ, and "B" shows that this is a linear pot. Also note the pin numbers on the connections. They don't really match the configuration on the pot. Sometimes circuit diagrams are drawn for the convenience of seeing the connections, with less attention paid to how the real-world component is laid out.

- Use jumper cables to make connections between the 10kΩ linear pot and the breadboard. First connect pin 3 of the pot to the voltage supply. This is shown as connection 1 in figure 10-38. You will need to insert some short pieces of wire into the breadboard for attaching the alligator clips.

Figure 10-38.

- Connect pin 1 (on the right) of the pot to the breadboard ground. This is connection 2 in the photo. Be careful to avoid touching the metal case of the pot with the jumper cables. Maintain a gap between the alligator clips and the case, as shown in figure 10-39.

- Turn the pot's shaft fully clockwise.

- Set your DMM for reading voltage.

- Use a jumper cable with alligator clips to connect the red probe of the DMM to pin 2 of the pot, as shown the photo. This is connection 3 in the photo.

- Use another jumper cable to connect the black DMM probe to the ground. This is connection 4 in the photo.

Figure 10-39.

- On the DMM, you should be seeing a voltage level that is the same as the supply voltage, about 9V.

- As you turn the pot's shaft counterclockwise, you should see a smooth decrease in the voltage level. The voltage should read zero when the pot is turned all the way down. This smooth decrease is what we expect from a linear taper.

- If you aren't seeing this variation in voltage, check that all of the connections are correct. In particular, check that there are no unintended connections (shorts) between the alligator clips where they are clipped onto the pins of the pot.

- Disconnect the pot from the breadboard and the DMM.

Audio Pot in a Circuit

This circuit is identical to the one you used for the linear pot, so you'll use the same procedure. The circuit diagram is the same also, because there is no difference in the pot symbol to show that a pot is linear or audio. The only difference is that the pot would be labeled **R1 – 10kA** on the diagram.

- Use jumper cables to make the connections between the 10kΩ audio pot and the breadboard: pin 3 of the pot connected to the voltage supply and pin 1 connected to the ground.

- Turn the pot's shaft fully clockwise.

- Set your DMM for reading voltage.

- Use jumper cables to connect the black DMM probe to the ground and the red DMM probe to pin 2 of the pot.

- On the DMM, you should be seeing a voltage level that is the same as the supply voltage, about 9V.

- As you turn the pot's shaft counterclockwise from full to half, you should see a sharp decrease in the voltage level, so that the level is down to about one-tenth of the supply voltage when the shaft is set to half. This decrease is what we expect from an audio taper pot.

- If you aren't seeing this variation in voltage, check that all of the connections are correct. In particular, check that there are no unintended connections (shorts) between the alligator clips where they are clipped onto the pins of the pot.

- Disconnect the pot and DMM.

This exercise has familiarized you with the voltage divider and the potentiometer. You've seen the voltage divider at work where it has reduced an input voltage by a particular percentage. You've also seen how a pot's value and its taper can be determined with a few simple tests.

11

Capacitors

Goal: When you have completed this chapter, you will be familiar with the capacitor's ability to store an electrical charge.

Objectives

- Understand the function of a capacitor.
- Recognize the symbols for polarized and nonpolarized capacitors.
- Identify a capacitor's value based on its markings.
- Observe a capacitor holding and releasing a charge.
- Build a circuit to observe the charging and discharging of a capacitor in series with a resistor, and draw a graph of these observations.
- Use these measurements to calculate the time constant.

The Property of Capacitance

Capacitance is defined as the ability to hold or store an electrical charge. In an equation, capacitance is represented by a capital letter **C**. This ability is exhibited by a component called a capacitor, which is designed to hold a particular amount of charge. The unit for measuring capacitance is the *farad*, named for Michael Faraday, which is abbreviated with an **F**. The farad is very large, and it is not a very handy unit to work with. Consequently, capacitance is usually measured in μF (microfarads), nF (nanofarads), and pF (pico-farads). A microfarad is one millionth of a farad, a nanofarad is one billionth of a farad, and a picofarad is one trillionth of a farad. Some multimeters have a setting for measuring capacitance, but most of them do not. There are also dedicated capacitance meters.

Capacitors

The structure of a capacitor is fairly simple: two plates of foil or other thin metal are separated by a narrow gap that is filled with a material called a *dielectric*: air, paper, or some other nonconductive material. If a battery or other voltage source is connected to one side and the other side is connected to the ground, the surface of the foil becomes charged: a positive charge accumulates on the surface connected to the voltage source, and a negative charge accumulates on the other surface, as shown in figure 11-1.

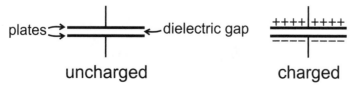

Figure 11-1.

The amount of charge that can be stored in a particular capacitor is proportional to the surface area of the plates, the type of dielectric, and how close together the plates are. Once we've made a metal-dielectric-metal sandwich, we can make it into stacks or roll it up into a cylinder. While smaller capacitors are often disc shaped or blob shaped, the really big capacitors are recognizable from their tubular shape. The structures and shapes of several types of capacitors are shown in figure 11-2.

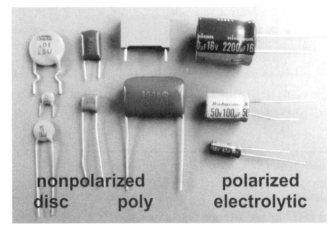

Figure 11-2.

Capacitor Symbols

There are a number of symbols for capacitors. They are all variants on the basic symbol, which is essentially a diagram of how a capacitor is constructed, with two plates separated by a gap of dielectric. These are shown in figure 11-3.

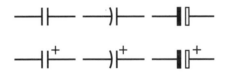

Figure 11-3.

Some capacitors are constructed with a solution of electrolytes (salts) soaking the dielectric, which etches the surface of a piece of foil and greatly increases the surface area. These electrolytic capacitors are polarized, so they can only be hooked up one way. As shown in figure 11-4, the negative terminal is the one that is marked. Just as with LEDs, the positive lead is the longer one. In the symbols, the positive terminal is the one that is marked. There are also tantalum capacitors, which are polarized, usually with the (+) terminal being marked. On circuit diagrams and circuit boards, all capacitors are labeled **C**, regardless of whether they are nonpolarized or polarized capacitors. The only indication of being polarized is a + sign in the symbol and on the circuit board.

Figure 11-4.

How Capacitors Charge

Before we start charging up the capacitor, the amount of charge on each plate is equal. When we connect the capacitor to a voltage source, current flows onto one plate. Its influence travels across the dielectric and forces an equal amount of current to flow off of the second plate. This follows Kirchhoff's current law, which says that the amount of current entering a point has to be equal to the current leaving the point. In this case, the amount of current entering the capacitor and being stored on one plate has to be balanced by an equal amount flowing off the other plate. It is this ability to let the electrons on one plate feel the electrons flowing onto the other plate that makes the material a dielectric. Not only must it be an insulator, but it must also be able to transmit the influence of charged particles from one plate to the other.

We now need to think about what is happening as the capacitor is charging. Current is flowing into one wire and flowing out of the other. From the outside, it looks as if the two wires are directly connected, and yet we know that they are not. When we start the charging process, the capacitor is acting like a conductor. This process of charging up the capacitor is shown in figure 11-5.

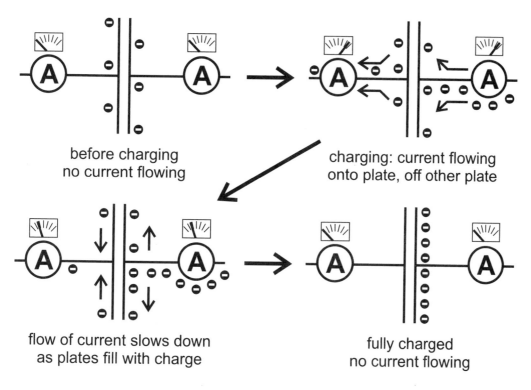

before charging
no current flowing

charging: current flowing
onto plate, off other plate

flow of current slows down
as plates fill with charge

fully charged
no current flowing

Figure 11-5.

As this charging continues, the charge builds up on one plate, while an equal amount of charge is driven off of the other plate. After a while there is less and less room for charge to keep building up, and the movement of current onto the plate slows down to a trickle and eventually comes to a halt. At the same time, the equal movement of current off of the other plate similarly slows down and stops. Now we have a charge built up on the capacitor, and we have no current flowing in or out. The capacitor is no longer acting like a conductor. Instead it is acting like an insulator. It is also holding a charge, like a battery. If we put a voltmeter across the two wires of the capacitor, we will see the same voltage as the voltage source. If we now open a path between the two sides of the capacitor, the charge that is held there will make its way to the other side, and the stored energy will be used up.

How a capacitor acts with AC voltage is of particular interest. We saw earlier that when we first apply a voltage across the capacitor and it charges up, it's acting like a conductor. If we now switch the voltage source around, with positive turned to negative and negative turned to positive, the current through the capacitor will also change direction. The capacitor hasn't had a chance to build up a charge yet, so it's still acting like a conductor. If we keep switching the voltage source back and forth, the current will continue to switch back and forth. Only if we slow down the speed of switching will we see the charge build up further and the resultant reduction of current.

What this means for us is that a capacitor acts differently with DC and AC, and even acts differently at different frequencies. At high frequency, the capacitor is essentially invisible, acting just like a piece of wire. As the frequency lowers, the capacitor becomes more restrictive, until at DC it completely blocks the current.

Capacitors are used in two ways in a circuit. One function is to block DC current. After the initial charging from a DC source, a capacitor doesn't allow any current to flow through it. At first glance this doesn't look very useful. After all, why should we use a component to prevent current from flowing if we can just open the circuit? Where this ability to block DC turns out to be useful is to separate AC signal from DC. One example is the signal coming from a condenser microphone, which requires +48V phantom power. The preamp is expecting a very small AC signal, and +48V is much larger than any signal we would ever ask it to amplify. At the same time, we must have phantom power to run the head amplifier in the mic and to charge up the capsule (more on this in a moment). We can use a filtering capacitor to separate the small AC signal from the much larger DC voltage.

Capacitors are also used with resistors to make filters. You'll see a brief explanation of this in a later chapter, when we look at the inductor. An important factor in this ability to make a filter is how fast the capacitor charges up to its maximum voltage. How long it takes a capacitor to charge up depends on its capacitance (measured in farads) and on the resistance the current encounters on its journey into the capacitor (measured in ohms). Together the capacitance and resistance give us a *time constant*, which is measured in seconds. The symbol for the RC time constant is τ, which is the lowercase Greek letter *tau* (rhymes with "cow"). The equation for τ is simply the capacitance and the resistance multiplied together.

$$\tau = R \times C$$

The time constant τ is a measurement of how long it takes to charge up the capacitor to 63% of its total possible charge. This constant is independent of the voltage applied to the capacitor and resistor. Since τ is the product of the resistance and the capacitance, voltage is not part of the calculations. For example, we can find τ for a 100μF capacitor used with a 47kΩ resistor.

Quantities: We list the quantities that we know, and the one quantity that we're going to solve for. If there are any prefixes, we remove them.

C = 100μF = 0.0001F
R = 47kΩ = 47,000Ω
τ = ???

Equation: Identify the equation we need, and rearrange this equation so we can solve for the quantity that we don't know.

$\tau = R \times C$

Solve: Combine the quantities that you know with the equation, and solve the math.

$\tau = 47,000\Omega \times 0.0001F$

$\tau = 4.7 \text{ sec}$

Prefix: Add a prefix if one is needed. Here we don't need a prefix, because the answer is already between 1 and 999.

$\tau = 4.7 \text{ sec}$

We will examine and observe the time constant τ further in the hands-on exercise.

Capacitor Value Codes

Just as resistors are marked with their values, so are capacitors. Unlike resistors, capacitors are marked with numbers and letters instead of color-coded bands. While the code for resistors is standardized, the codes for capacitors are somewhat less uniform. There are four distinct systems of capacitor value marking commonly in use today. Just to confuse us, there may be some other letters and numbers printed on the capacitor along with the value code. These indicate the manufacturer's date code, the lot number, or other quality-control information. We can safely ignore this part of the label. The trick is teasing out which part is important and which is not. Several of these codes include a tolerance code, as shown in table 11-1.

These codes can be really confusing, because in other situations, "M" stands for mega, "k" stands for kilo, and "F" stands for farad. It would have been nice if the industry could have chosen other letters here, but they didn't. We just have to remember that when reading capacitors,

Capacitor tolerance codes	
code	tolerance
F	± 1%
G	± 2%
H	± 3%
J	± 5%
K	± 10%
M	± 20%
N	± 30%
Z	+80%, - 20%

Table 11-1.

these are tolerance codes, not prefixes or units. After a bit of practice, it becomes automatic. The tolerance of a capacitor is rarely a critical concern in audio electronics, except for when a particular value is needed for a filter, so most of the time we can ignore the tolerance.

The Three-Number Code

The most common code has some similarity to the resistor code, in that it uses three digits. The unit of measurement for this code is always picofarads (pF). The first two numbers are the digits; the third is the number of zeroes that follows them. The first digit can't be a zero. A letter code follows this, indicating the tolerance. After that, either on the same line or the next line is the working voltage rating, which may or may not have a letter "V" after it to indicate volts. The voltage rating is also called the maximum working voltage, which is the highest voltage to which you would want to expose the capacitor.

There is a little wrinkle to this code. As described, it doesn't allow for values of less than 10pF: a marking of 100 would be "10 with zero zeroes." Just as with the resistor code, we need a "divide by" code. A "9" in the multiplier means "divide by 10." An "8" in the multiplier means "divide by 100." It's all right to use 8 and 9 for this purpose, because any capacitor with two digits and eight zeroes would be impractically large. (The technical standard also says that 6 is not used as a multiplier, but the industry seems to be using it anyway: I have film capacitors labeled **106J** for 10μF±5%.) For example, the marking shown in figure 11-6 denotes a component that has a capacitance of 0.047μF, with a tolerance of 10%.

Just as with the resistor code, we change 4-7-3 to 47,000pF, then move the decimal point over six places to get μF. The general rule for resistors is not to have leading zeros, but with capacitors it's much more acceptable to have a designation with a leading zero, such as 0.047μF, because capacitors are usually labeled in either μF or pF. The use of nF is still technically correct and is seen on some schematics, but it is not as common yet.

Figure 11-6.

The Value in Microfarads

Another identification code scheme is to print the number of microfarads, followed by the tolerance code, followed by the maximum voltage. This code may be printed on one or two lines. As before, there doesn't have to be a V to indicate voltage, as it's understood that the second number is the voltage rating. This marking scheme is shown in figure 11-7, where a label of **.02J400** denotes a component that has a capacitance of 0.02μF with a tolerance of ±5%, rated at 400V.

Figure 11-7.

The Prefix as the Decimal Point

A similar code uses the prefix (μ, n, or p) as the decimal point. This creates fewer errors caused by the decimal point not being printed properly or just being too small to see. It also allows all the codes to be written with only three characters. For example, 4.7nF is written as **4n7**. 47nF is written as **47n**. 470nF is equal to 0.47μF, so it is written as **μ47**. An example of this is seen in figure 11-8, which shows a component that has a capacitance of 47nF, with a tolerance of 10%, rated at 63V.

Figure 11-8.

Direct Value Reading

The easiest labeling scheme is the one used on electrolytic capacitors. Here, the component's value is printed directly on the capacitor, along with its maximum working voltage. This is shown in figure 11-9. This component has a capacitance of 2200μF, rated at 35V. The other marking on this capacitor is the polarity. The negative lead of the capacitor is indicated by a strip of minus signs all along that side of the can. The positive lead is longer than the negative lead (until you solder it in and clip off the excess).

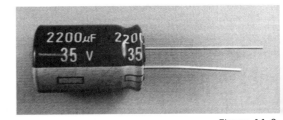

Figure 11-9.

SMT Capacitors

We learned earlier about Surface Mount Technology (SMT) resistors. There are some SMT capacitors that have no markings at all because they are assembled by robotic machinery that doesn't need to read the value; it just needs to know which roll the component comes from. Working on this equipment and replacing its capacitors requires a capacitance meter, because all of the capacitors are just little tan-colored blocks. Working on SMT circuit boards requires a steady hand, a very fine-tipped soldering iron, and either a stereo microscope or extraordinarily good eyesight. Some typical SMT capacitors are shown in figure 11-10, with the edge of a penny at the top for size comparison.

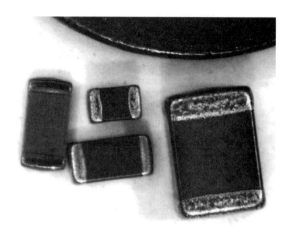

Figure 11-10.

Electrolytic Capacitors—the Good and the Bad

Electrolytic capacitors fill a certain niche in the electronics world. In their physical size, they hold a lot more charge than the same-sized film capacitor. For example, see figure 11-11 for a view of two 10µF capacitors. The big blob is a polypropylene film capacitor. The little can is an electrolytic capacitor. It's called electrolytic because of its construction, where a piece of foil is wrapped up in contact with a piece of paper soaked in a solution of electrolytes, essentially a salt solution. This electrolyte acts as the second electrode.

Figure 11-11.

Electrolytic capacitors have the advantage of size and cost, but they need a little extra care. Almost all of them are polarized, meaning that the (+) lead must be closer to the positive voltage, and the (–) lead must be closer to the negative voltage. If the capacitor gets turned around, the current flows the wrong way through the electrolytic solution, with disastrous consequences. If the backwards voltage goes on for too long or too high, the solution overheats and eventually boils off, exploding the capacitor. The remains of such an incident are shown in figure 11-12.

Once the capacitor is installed properly on the circuit board, it's less likely that it will get damaged in this way. Unfortunately the electrolytic solution doesn't last forever. Sooner or later the capacitor dries up and doesn't work properly anymore, whether the equipment is in use or not. The length of time it takes for capacitors to degrade and the determination as to when they should be replaced are matters of some debate in the audio electronics world. Some people in the industry say that electrolytic capacitors only last 10 or 15 years. Others say that there is 40-year-old equipment still working that is running on the original capacitors.

Figure 11-12.

Hot Melt Glue

There is one other aspect of the electrolytic capacitor (and other components) that bears mentioning here, which is the use of hot melt glue to stabilize the components on the circuit board. It would make sense that pushing the can of an electrolytic capacitor as far as it will go into its holes on the circuit board and soldering it into place should be the last time you need to worry about it until the electrolyte solution dries out in a decade or two. Under most circumstances this would be true. There is an exception, which is when the circuit board and its capacitors are subject to vibration. Consider the big capacitors in the circuit board of a bass head amplifier. Night after night it sits on top of a bass cab onstage being vibrated back and forth. Even though there isn't much of a gap between the bottom of the capacitor and the circuit board, the capacitor is still subject to inertia and is able to rock a little bit. This eventually weakens the connection of the leads inside the capacitor, and it only works intermittently or even just breaks off completely. This is shown in figure 11-13.

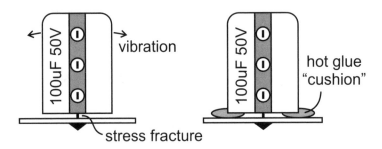

Figure 11-13.

The solution is to put a little cushion of hot melt glue at the base of the capacitor to stabilize it. When you open up a piece of gear and see little festoons of goo around the capacitors, that's what you are looking at. To soften and remove it, you can use an old soldering pencil, a heat gun, or even a hair dryer. Don't use your good soldering iron tip for this. It smells nasty, and the tip will need to be replaced because it won't hold solder properly anymore. (This is a good reason to keep a couple of old tips around, because they will still work for glue removal even after they don't work for soldering.)

The Condenser Microphone

A highly specialized capacitor used in the audio environment is the condenser microphone. The British name for capacitor is *condenser*, and the name has stuck to the microphone. In this microphone design, there is a rigid plate called the *backplate* and a moveable plate called the *diaphragm*. A tiny air gap separates the two plates. Earlier we said that air can act as a dielectric, the insulator between the plates of a capacitor. This sandwich of backplate and air and diaphragm can act as a capacitor. This assembly is the condenser microphone capsule, as shown in figure 11-14. A high voltage (typically about 300V) is placed across the plates to charge them up. When air pressure from a sound wave presses the diaphragm inward, this changes the thickness of the dielectric, and the capacitor can

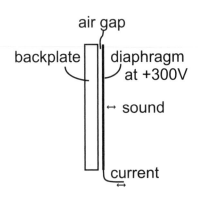

Figure 11-14.

hold more charge. Current flows onto the capacitor. The diaphragm returns to its original place, and current flows back off, because the capacitor can't hold as much charge anymore. Changes in air pressure (i.e., sound waves) are transformed into the movement of electrons.

Intentional and Unintentional Capacitance

Capacitors are essential to the audio circuit and are used in many ways. One common use is to remove ripple and hum. Another use is to remove the DC component from an AC signal. Capacitors are used as filters when combined with resistors and inductors. The quality of capacitance, the ability to hold a charge, is also a big factor in other parts of electronics that don't involve a manufactured capacitance. For example, cables exhibit the ability to hold a certain amount of charge, even though they aren't intended to act as capacitors. It is this capacitance that interacts with the impedances of the source and the input to act as a filter and cause the loss of high-frequency signal, especially in long cable runs. Whether we're looking directly at the components that are manufactured specifically to hold the charge or the materials that do so unintentionally, we can see that capacitance is an important concept to understand in audio electronics.

Hands-On Practice

The goal of this exercise is to familiarize you with the functioning of a capacitor. We'll use several different approaches to look at how a capacitor charges up, holds a charge, and discharges. You'll learn to identify capacitors from their coded markings.

Materials

These are the materials you will need for these exercises.

- Digital multimeter (DMM)
- Breadboard
- 9V battery and battery clip with wires
- 1.0kΩ resistor
- 10kΩ resistor
- 22kΩ resistor
- 100µF electrolytic capacitors
- 1000µF electrolytic capacitor
- LED
- Stopwatch or watch with a second hand
- 2 jumper cables with alligator clips
- Solid insulated wire (20, 22, or 24 gauge)
- Optional: grab bag of capacitors
- Optional: dead piece of audio gear

Charging and Discharging a Capacitor

In this section, we will be examining some of the charging and discharging behaviors of capacitors.

- Set up your breadboard with a 9V battery.

- Use a piece of wire to connect one of the upper component strips to the ground bus. This component strip is now considered to be part of the ground. (The only reason for doing this is that most capacitor leads are too short to stretch conveniently from the voltage supply bus to the ground.)

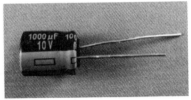

Figure 11-15.

- Locate the 1000μF electrolytic capacitor, as shown in figure 11-15. This capacitor is polarized, meaning that it can only be used in one direction. Note that it has one lead longer than the other, which identifies it as the (+) lead. Also notice that the side of the capacitor cylinder next to the shorter lead is marked with a row of minus signs, identifying it as the (−) lead. This is the other indicator of the polarity.

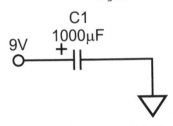

Figure 11-16.

- Construct the circuit shown in figure 11-16. This is a pretty small circuit, but there is still something to pay attention to: the longer (+) lead must be plugged into the voltage supply bus, while the shorter (−) lead gets plugged into the ground.

- The completed circuit is shown in figure 11-17.

- Measure the voltage across the leads of the capacitor, with the red probe on the (+) lead and the black probe on the (−) lead. The voltage should be 9V, the same as your battery.

- Remove the capacitor and turn it upside down.

- Carefully move the leads together until they just touch. Be prepared for a surprise: a spark will jump from one to the other when they get close enough. It may not be really dramatic the first time, so recharge the capacitor and try it again. Charge and discharge the capacitor as many times as you like. (This will eventually run down the battery.)

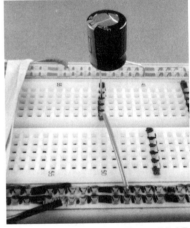

Figure 11-17.

- Charge the capacitor and remove it from the breadboard.

- This time, plug it into two unused component strips on the breadboard.

- Measure the voltage across the leads of the capacitor, as before. This is shown in figure 11-18.

- The voltage should be close to the 9V that you measured previously, but it will not be quite stable. We expect to see the voltage dropping slowly. This gradual reduction is because the dielectric doesn't form a perfect barrier to the charge held on the two plates. A little bit of current manages to leak across, and the voltage is slowly reduced.

Figure 11-18.

- Charge the capacitor again, remove it, and plug it into two unused component strips again.

- Use the resistor color code to identify a 1.0kΩ resistor.

- This time we will use the charged-up capacitor to do a little work for us. Assemble the circuit shown in figure 11-19. Be careful to pay attention to the polarities of the LED and the capacitor.

- When you assemble the circuit, the electrons stored in the capacitor will flow through the resistor and LED, making the LED light up briefly. The LED starts off lighting strongly, but it quickly fades as the charge in the capacitor is depleted.

- You can repeat the cycle of charging up the capacitor with the voltage supply, removing it, and placing it in this circuit to discharge it. This cycle gives us an idea of how much charge is held in the capacitor, compared with how much the battery is generating.

- Clear the breadboard for building later circuits.

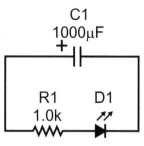

Figure 11-19.

Observing the Capacitor Charging Rate

Now that the capacitor is fully charged, we can also watch the capacitor discharge. How long it takes a capacitor to discharge is also governed by the time constant τ of the capacitor and the resistance.

- You will be collecting 13 points of data and using them to draw a small graph. Just as for the potentiometer resistance graphs, you can collect the data in a table and draw it on regular paper following the format of the data table and blank graph shown below, or you can print out the blank data table and blank graph from the supplemental materials on the DVD.

- To collect the data of the capacitor charging, you will build a circuit where you will be able to constantly monitor the voltage of the capacitor. The beginnings of this circuit are shown in figure 11-20.

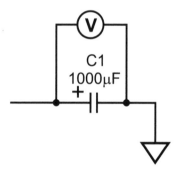

Figure 11-20.

- Here we see a 1000µF electrolytic capacitor with its (−) lead tied to the ground and a voltmeter attached across it, as shown in figure 11-21.

- Use the resistor color code to identify a 22kΩ resistor and a 10kΩ resistor.

- Measure the actual resistances of these two resistors and write them down.

- Assemble the next part of the circuit, as shown in figure 11-22. The 22kΩ resistor is connected to the voltage supply and to an unused component strip. The 10kΩ resistor is connected to the ground and to an unused component strip. A piece of wire connects the (+) lead of the capacitor to the ground. This circuit will be altered to allow the capacitor to charge by way of the 22kΩ resistor while you monitor and record the voltage level at 10-second intervals. Later the capacitor will be allowed to discharge through the 10kΩ resistor while you monitor and record the voltage level at 10-second intervals.

Figure 11-21.

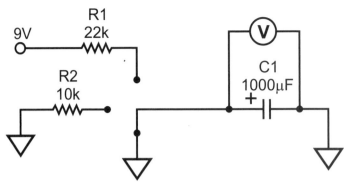

Figure 11-22.

- The voltage level across the capacitor should be zero, as shown on the DMM. This circuit is shown in figure 11-23. Unplug the piece of wire from the ground. When you do this, the voltage across the capacitor may rise a few tenths of a volt. This is an expected result. Even though the capacitor has been completely discharged by grounding both leads, the dielectric holds on to a little bit of "memory" of being charged up.

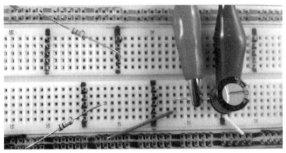

Figure 11-23.

- The circuit for charging the capacitor is shown in figure 11-24. Do not connect the capacitor to the resistor yet.

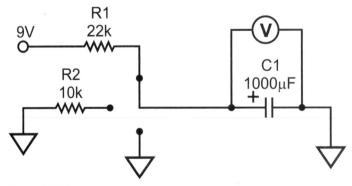

Figure 11-24.

- Wait to connect the circuit until you are prepared with your stopwatch or timer, something to write with, and a piece of paper with the data table shown in table 11-2.

charging		discharging	
time	voltage	time	voltage
0:00	0	0:00	
0:10		0:10	
0:20		0:20	
0:30		0:30	
0:40		0:40	
0:50		0:50	
1:00		1:00	
1:10		1:10	
1:20		1:20	
1:30		1:30	
1:40		1:40	
1:50		1:50	
2:00		2:00	

Table 11-2.

- Start your stopwatch at exactly the same time that you connect the wire to the 22kΩ resistor, as shown in figure 11-25. After that, at every 10-second time point, look at the voltage shown on the DMM display, and write it down on the "charging" column of the table. You should see a fairly fast rise in the voltage at first, and then it will continue to increase more slowly.

Figure 11-25.

- When 2 minutes have passed and the table of voltages is completely filled in, leave the capacitor plugged in. Turn off and reset your stopwatch or timer.

- The graph format is shown in the chart in figure 11-26.

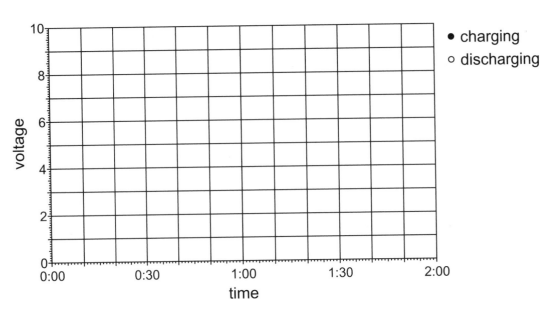

Figure 11-26.

- Graph the voltages on the chart with closed circles. There should be one data point on each vertical line.
- Connect the data points together in a smooth curve. An example set of data and graph is shown at the end of the exercise.
- Measure the voltage of the battery supply and write it down on the sheet of paper.
- Multiply this voltage by 0.632. The voltage you calculate by doing so is the charge that will be on the capacitor after one time constant τ.
- Locate this voltage level on the left axis of the graph, and draw a horizontal line across the graph to mark this voltage level.
- Where this line crosses the curve of the voltage, draw a vertical line down to the time axis along the bottom. This is shown in the sample graph at the end of the exercise.
- Estimate and write down what time this vertical line represents, as shown in the sample.
- Use the capacitance of 1000μF and the actual resistance of the 22kΩ resistor to calculate the time constant τ. The equation is $\tau = R \times C$. Use the four steps (Quantities, Equation, Solve, Prefix) for this calculation. It is essential that the prefixes be removed before making the calculation, as shown in the example in the chapter text.
- You should get an answer similar to the time shown by the vertical line of the graph. This is the time constant for the combination of this capacitor and this resistor. After five time constants, the capacitor is considered fully charged. Your observed answer may be off by a few seconds, just as seen in the example at the end of the exercise. This is to be expected, since there is some variation in the actual capacitance of the component.

Here we are working with a combination of capacitor and resistor that has a time constant of many seconds. In the audio world, we're usually working with capacitances and resistances that give time constants of milliseconds or even microseconds, but the process is exactly the same. When we apply an external voltage, the voltage level in the capacitor rises at a rate defined by the time constant of that capacitor and that resistance.

Observing the Capacitor Discharging Rate

Now that the capacitor is fully charged, we can also watch the capacitor discharge. How long it takes a capacitor to discharge is also governed by the time constant τ of the capacitor and the resistance.

As for the charging-rate exercise, you will be collecting 13 points of data and using them to draw a small graph. You can collect the data in a table and draw it on regular paper following the format of the data table and blank graph.

• From the previous exercise, you have the circuit assembled and ready to go.

• The capacitor should be fully charged with about 9V. Write down this voltage as the first data point at 0:00.

• The discharging circuit is shown in figure 11-27. Do not build this circuit yet.

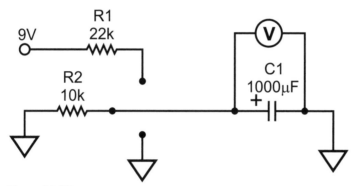

Figure 11-27.

• A photo of the discharging circuit is shown in figure 11-28.

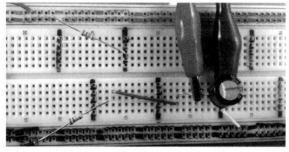

Figure 11-28.

• Disconnect the piece of wire from the 22kΩ resistor.

• Start your stopwatch at exactly the same time that you connect the wire to the 10kΩ resistor that is connected to the ground. After that, at every 10-second time point, look at the voltage shown on the DMM display and write it down on the table. You should see a fairly fast drop in the voltage at first, and then it will decrease more slowly after that.

- When 2 minutes have passed and the table of voltages is completely filled in, leave the capacitor plugged in. Turn off and reset your stopwatch or timer.

- Graph the voltages on the chart with open circles. There should be one new data point on each vertical line.

- Connect the data points together in a smooth curve. An example set of data and graph is shown at the end of the exercise.

- Multiply the starting voltage by 0.632. The voltage you get is the charge that was lost by the capacitor after one time constant τ.

- Subtract this voltage from the starting voltage at time 0:00. The voltage you calculate here is how much charge is left on the capacitor after one time constant τ.

- Locate this voltage level on the left axis of the graph, and draw a horizontal line across the graph to mark this voltage level.

- Where this line crosses the curve of the voltage, draw a vertical line down to the time axis along the bottom. This is shown in the sample graph at the end of the exercise.

- Estimate and write down what time this vertical line represents, as shown in the sample.

- Use the capacitance of 1000μF and the actual resistance of the 10kΩ resistor to calculate the time constant τ. The equation is τ = R × C. Use the four steps (Quantities, Equation, Solve, Prefix) for this calculation. Remember that the prefixes must be removed before making the calculation, as shown in the example in the chapter text.

- You should get an answer similar to the time shown by the vertical line of the graph. Just as for the capacitor-charging exercise, this is the time constant for the combination of this capacitor and this resistor. Again, you may not see the exact time constant you calculated, but the answers will be pretty close. After five time constants of discharging, the capacitor is considered fully discharged.

This set of exercises has given you experience with identifying capacitors and watching them charge and discharge. You've calculated the time constant τ, which governs how fast a particular capacitor and resistor combination allow a capacitor to charge up and discharge. You've also seen how important it is to check that an electrolytic capacitor is placed in the correct orientation in a circuit.

Determining Capacitor Markings

In this part of the exercise, you will write out the capacitor value markings you would see on a capacitor of a particular value. Some examples are given, and the answers are given at the very end of the exercise. Until we get experienced with this process, we're not going to worry about the markings for the tolerance or the working voltage.

- Example 1: Determine the value markings for a 0.022μF capacitor, using the three-number code, the microfarad code, and the prefix-as-decimal code.
  - Three-number code: For the three-number code, we need the value expressed in pF (picofarads). At the moment, the value is in μF (microfarads). To get to the value in pF, move the decimal six places to the right (multiply by 1 million). 0.022μF becomes 22,000pF. The digit-digit-multiplier code for 22,000 is 2-2-3, so we're looking for a capacitor labeled **223**.
  - Microfarad code: 0.022μF is already written out in microfarads, so we're simply looking for a capacitor labeled **0.022**.

- Prefix-as-decimal code: We need to move the decimal to wherever we can just use two numbers and the prefix. Move the decimal over three places to the right (multiply by 1000) to change 0.022μF to 22nF. Without the F, that's it: we're looking for a capacitor labeled **22n.**

- Example 2: determine the value markings for a 5.6nF capacitor, using the three-number code, the microfarad code, and the prefix-as-decimal code.
  - Three-number code: For the three-number code, we need the value expressed in pF (picofarads). At the moment, the value is in nF (nanofarads). To get to pF, move the decimal three places to the right (multiply by 1000). 5.6nF becomes 5600pF. The digit-digit-multiplier code for 5600 is 5-6-2, so we're looking for a capacitor labeled **562.**
  - Microfarad code: To use the microfarad code, the value in nanofarads needs to be rewritten in microfarads. To get there, move the decimal three places to the left (divide by 1000). 5.6nF becomes 0.0056μF. We're looking for a capacitor labeled **0.0056.**
  - Prefix-as-decimal code: We don't need to move the decimal anywhere; just substitute the decimal and remove the F. 5.6nF becomes a label of **5n6.**

- Example 3: determine the value markings for a 680pF capacitor, using the three-number code, the microfarad code, and the prefix-as-decimal code.
  - 680pF = 681, 0.00068, n68

- Example 4: Determine the value markings for a 82pF capacitor.

- Example 5: Determine the value markings for a 3.9μF capacitor.

- Example 6: Determine the value markings for a 1.8nF capacitor.

- Example 7: Determine the value markings for a 150nF capacitor.

- Example 8: Determine the value markings for a 470pF capacitor.

- Example 9: Determine the value markings for a 10μF capacitor.

- Example 10: Determine the value markings for a 0.027μF capacitor.

Reading Capacitor Values

In this part of the exercise, you will look at some photos of capacitors to determine their values. Some examples are given, and the answers are provided at the very end of the exercise.

- Example 11: Figure 11-29.

- Example 11 shows a capacitor labeled **(M)275K 250E.** We interpret this as 2.7μF, 10% tolerance, rated at 250V. (M) is a manufacturer's logo. We can ignore the E.

Figure 11-29.

- Example 12: Figure 11-30.

- Example 12 shows the close-up view of a capacitor labeled **68n G 63V BC 0246416.** We interpret this as 68nF, 2% tolerance, rated at 63V. The rest is manufacturing information.

Figure 11-30.

- Example 13: Figure 11-31.
- Example 13 shows a capacitor labeled **BC C0 151J 100V.** We interpret this as 150pF, 5% tolerance, rated at 100V. C0 is the dielectric. BC is the manufacturer.

Figure 11-31.

- Example 14: Figure 11-32.

Figure 11-32.

- Example 15: Figure 11-33.

Figure 11-33.

- Example 16: Figure 11-34

Figure 11-34.

- Example 17: Figure 11-35.

Figure 11-35.

- Example 18: Figure 11-36.

Figure 11-36.

- Example 19: Figure 11-37.

Figure 11-37.

- Example 20: Figure 11-38.

Figure 11-38.

Grab Bag Measurements and Dead Gear Measurements

- If you have a grab bag of capacitors, take out an assortment and practice reading their values.
- Don't be surprised to find a few that are difficult to interpret. The electronics industry has slowly settled on some labeling standards for capacitors, but there are so many "standard" methods that it may not be simple to figure out which one is being used. Don't expect to be able to identify them all at first glance.
- If you have a DMM that reads capacitance, try it out on these capacitors.
- If you have a dead piece of audio gear, open it up and look around for capacitors. You can expect to find quite a variety. There will be discs, boxes, blobs, slabs, and cylinders of many sizes.
- If you can see the labels, try reading their values.
- Check the tops of the cans of the electrolytic capacitors for signs of swelling or leakage. If there is a crusty residue on the top of the can, you know the capacitor has had some of the electrolyte solution escape.

Answers

- Example 4: 82pF = 820, 0.000082, 82p.
- Example 5: 3.9µF = 395, 3.9, 3u9.
- Example 6: 1.8nF = 182, 0.0018, 1n8.
- Example 7: 150nF = 154, 0.15, u15.
- Example 8: 470pF = 471, 0.00047, n47.
- Example 9: 10µF = 106, 10, 10u.
- Example 10: 0.027µF = 273, 0.027, 27n.
- Example 14: **135 01 16v 1500µF** = 1500µF, rated at 16V, electrolytic as shown by the (–) sign on the band.
- Example 15: **(HJC) 2.2µ 250MPP** = 2.2µF, 20% tolerance, rated at 250V.
- Example 16: **01 Z5U** = 0.01µF, +80%, -20% tolerance
- Example 17, as shown in Figure 11-26: **MM125K 2G S** = 1.2µF, 10% tolerance.
- Example 18, as shown in Figure 11-27: **220µF 35V** = 220µF, rated at 35V, electrolytic as shown by the (–) sign on the band below the label.
- Example 19, as shown in Figure 11-28: **P104G 50V (M).F.** = 0.1µF, 2% tolerance, rated at 50V.
- Example 20, as shown in Figure 11-29: **47nK MMK BV9 63-** = 47nF, 10% tolerance, rated at 63V.
- An example of charging and discharging data is shown in table 11-3.

charging		discharging	
time	voltage	time	voltage
0:00	0	0:00	8.95
0:10	2.89	0:10	3.68
0:20	4.75	0:20	1.66
0:30	5.89	0:30	0.84
0:40	6.73	0:40	0.45
0:50	7.31	0:50	0.25
1:00	7.73	1:00	0.15
1:10	8.06	1:10	0.10
1:20	8.29	1:20	0.07
1:30	8.48	1:30	0.05
1:40	8.61	1:40	0.04
1:50	8.71	1:50	0.04
2:00	8.79	2:00	0.03

Table 11-3.

- Charging: An example graph is shown in figure 11-39.

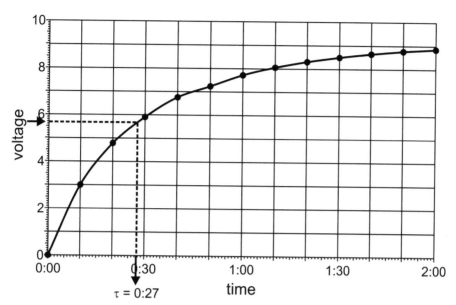

8.95V × 0.632 = 5.66V, gives observed τ = 27 sec
calculated τ = 21,700Ω × 0.001F = 21.7 sec

Figure 11-39.

- Discharging: An example graph is shown in figure 11-40.

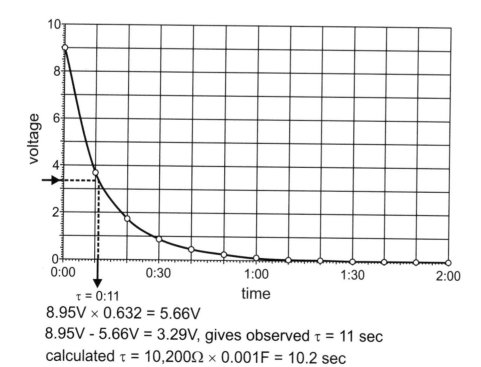

8.95V × 0.632 = 5.66V
8.95V - 5.66V = 3.29V, gives observed τ = 11 sec
calculated τ = 10,200Ω × 0.001F = 10.2 sec

Figure 11-40.

12

Inductors and Filters

Goal: When you have completed this chapter, you will have a working knowledge of the inductor's ability to store energy in a magnetic field.

Objectives

- Understand the function of an inductor.
- Recognize the symbols for air-core and iron-core inductors.
- Identify low-pass and high-pass filters in circuit diagrams.

Inductors

An *inductor* is an electronic component that is made of wire wrapped in a loop, a tube, or a torus shape (like a donut). It stores energy in the form of a magnetic field. When a current is running through the inductor, the field is created in the space surrounding the coil. The schematic symbols for inductors are shown in figure 12-1.

On the left is an air-core inductor; on the right is an iron-core inductor. The amount of energy that can be stored in an iron-core inductor is much higher. Some inductors are shown in figure 12-2.

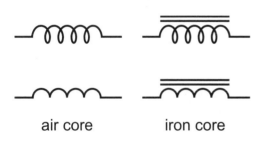

air core   iron core

Figure 12-1.

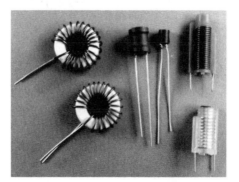

Figure 12-2.

The ability to store energy in a magnetic field is called *inductance*. This term is used to specify the amount of energy that can be stored by an inductor. The symbol for inductance or an inductor is an **L,** and the standard unit of inductance is the *henry*, which is abbreviated **H.** The L for inductance was chosen to honor the physicist Heinrich Lenz, who pioneered electromagnetism in the mid-1800s. The henry is named after Joseph Henry, who invented the electromechanical relay that became the basis of the telegraph. The henry is a very large unit, so the unit typically used for measuring inductors is the millihenry (mH) or the microhenry (μH). In circuit diagrams, inductors are labeled with an **L.**

How Inductance Works

Before current flows through an inductor, there is no magnetic field in or around it. When a voltage is first applied across an inductor, current tries to flow through, but it is very difficult because there is no magnetic field yet. The small current that does flow creates a magnetic field, which in turn makes it easier for current to flow. Current flow increases, and the magnetic field continues to build until it reaches the limit of energy storage defined by the component's inductance. When we started, the inductor was restricting how much current was flowing through it, so it was acting like a resistor. Once it stores its maximum amount of energy in a magnetic field around the inductor, it isn't restricting current at all, so it's acting as if it has no resistance. As long as current continues to flow, the magnetic field just stays there, and the inductor acts like a piece of wire with no resistance. This is shown in figure 12-3.

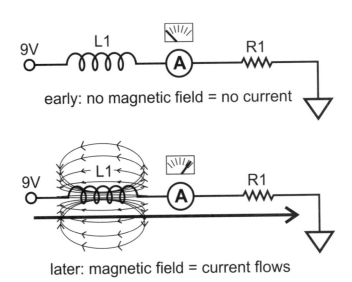

Figure 12-3.

In a normal circuit, if we disconnect the voltage source, any current in the circuit stops immediately; there is nothing to keep pushing or pulling it through. In a circuit with an inductor, the situation is different. When we first applied voltage, current didn't start flowing right away, because the magnetic field needed to be built up first. In the same way, when we remove the voltage, the current doesn't stop flowing right away. This is because of the energy stored in the magnetic field, which keeps pulling the current along for a while. The larger the inductor, the more energy is stored, and the longer the current keeps flowing.

In this way, the magnetic field of an inductor provides a kind of inertia. Sir Isaac Newton studied and described inertia, which can be briefly described in the following way: "An object in motion tends to remain in motion, and an object at rest tends to remain at rest." The magnetic field is like an object in in11 motion. It has inertia, so it keeps moving even after it is no longer being pushed.

This is easier to envision if we think in terms of physical inertia. Let's say we have a team of people pushing a car with the engine off. From the moment they start pushing, it takes a while to get the car to where it's rolling along smoothly. It takes time to overcome the inertia of the car, because the car wants to remain at rest. However, once it is moving, it's not going to come to a dead stop the instant they stop pushing on it. It will keep rolling until it runs out of energy. We stored energy in the car's movement. The situation is the same with DC voltage and an inductor. When we connect the voltage, it stores energy and current flows. When we disconnect the voltage, current keeps flowing because of the stored energy.

Now let's consider what happens with AC voltage, where the current direction is being changed with each cycle of the wave. Initially, the voltage is positive, so the inductor starts to develop the magnetic field. At this point, the apparent resistance of the inductor is high, because it won't let through very much current until the magnetic field is created. If the voltage is suddenly changed to negative, the inductor pushes back against the current coming through; what magnetic field it has managed to develop is still pulling the current along in one direction, while the voltage source is trying to push it in the other direction. The existing magnetic field needs to be used up before a new one in the opposite can be developed. Given enough time, the inductor can do this, but with an AC source it will just be a moment or less before the voltage cycles back to being positive.

For envisioning an inductor with AC voltage, we'll go back to our car, but this time with two teams of people to push it in opposite directions. If we tell team 1 to push for 5 seconds, then tell team 2 to push for 5 seconds, then tell team 1 to push again for 5 seconds, the car will not move at all. Even though each team is putting energy into pushing the car for 5 seconds at a time, no team ever gets to overcome the resting inertia of the car, so it never starts moving. Even if it begins moving a just a little, the other team pushes against it and pushes it back an equal amount.

The consequence is that inductors let DC and low-frequency AC signals pass through them. They block high-frequency AC, because the magnetic field never has a chance to develop. Because of this inertial effect, an inductor can be used as a filter to block high-frequency signals. At a low frequency, the voltage continues long enough in one direction to overcome the inertia of the field in the inductor, and current passes through the inductor. There is a lag between when the voltage is first applied and when the current flows, but there's also a lag between when the voltage is turned off and when the current stops. No energy is lost, it is just stored in the magnetic field and harvested later. The signal is passed through as if the inductor has very low resistance. At a high frequency, the amount of time the voltage continues in a single direction is not long enough for it to outlast the inertia of the inductor, and the signal is blocked. The inductor acts like a very large resistor, one that is large enough to block the signal completely.

Filters and Impedance

This chapter and the previous one on capacitors are both discussing components that store energy and release it later. Inductors and capacitors both store energy, but they do it in different ways, and in consequence they act differently on DC and AC voltages. Figure 12-4 shows a chart of the effect of inductors and capacitors on DC and AC. This chart shows that capacitors and inductors act in opposite ways.

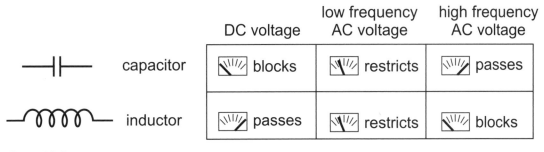

	DC voltage	low frequency AC voltage	high frequency AC voltage
capacitor	blocks	restricts	passes
inductor	passes	restricts	blocks

Figure 12-4.

We use these differences in how capacitors and inductors work to use them with resistors to build filters, which are essential parts of audio. We use low-pass, high-pass, peaking, and shelving filters in the equalization (EQ) process. We also incorporate them into audio gear to filter out frequencies that we don't want to amplify, such as radio frequency waves, and we also use filters in digital audio. These filters are made by combining resistors with capacitors (RC filters), by combining resistors with inductors (RL filters), and by combining all three together (RLC filters).

A *high-pass filter* allows through all signals that are above a certain frequency and reduces all frequencies below that frequency. The frequency where the filter starts to work is called the *cutoff frequency*, often shortened to f_c. The lower the frequency is below this cutoff, the more the signal is reduced. We call this "rolling off" the signal. In a simple filter, this reduction occurs at 6dB per octave. On a graph, we see this as a sloping line, so this reduction is called the *slope* of the filter. If a filter has a slope of –6dB per octave, we call it a *first-order filter*.

A *low-pass filter* is the opposite of a high-pass filter. It allows through all signals that are below the cutoff, and it rolls off all signals that are above the cutoff. As for the high-pass filter, the sloping line of a simple low-pass filter is –6dB per octave. This makes it a first-order filter as well.

The *order* of a filter is a measurement of the slope of the filter, and is completely independent of whether it is a low-pass or high-pass filter. A first-order filter has a slope of –6dB per octave. Later we will look at how we combine first-order filters to get sharper reduction of the signal. Figure 12-5 shows how to construct high-pass and low-pass first-order RC filters from resistors and capacitors.

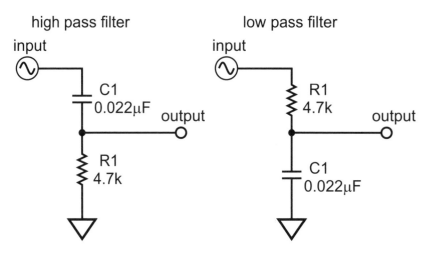

Figure 12-5.

The frequency response curves for these first-order filters are shown in figure 12-6.

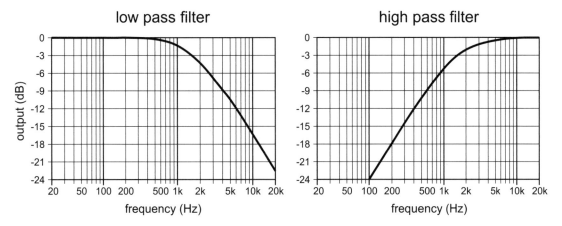

Figure 12-6.

The RC filter shown acts by blocking the low frequencies and letting the high frequencies through. To see how this works, we can think of the capacitor acting like a large resistor at low frequencies. This large resistance forms a voltage divider with the resistor of the filter. The capacitor is acting like R1 of the divider, and the resistor is acting like R2. With R1 large compared to R2, this configuration gives a low output at low frequencies. At high frequencies, the capacitor acts like a small resistor. This resistance is acting like R1 of a voltage divider, and the actual resistor is still acting like R2. This time, R2 dominates the divider, and the output is large. This is shown in figure 12-7.

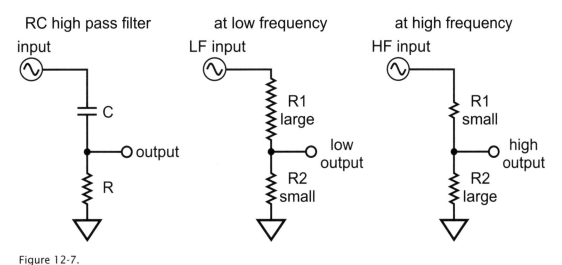

Figure 12-7.

Notice that the resistor, labeled as R2, hasn't changed in size, but at low frequencies it is labeled as being small and at high frequencies it is labeled as being large. These labels represent that its size is relative to the apparent resistance of the capacitor, which is large at low frequencies and small at high frequencies.

Here we see that we can use the voltage divider as a model to describe the action of the RC high-pass filter. The same model can be used for RC low-pass filters and for RL filters. If we delve into the math, the calculations that we do to find the cutoff frequency and the slope of the frequency roll-off are all based on the voltage divider equation. Even if we don't do the math, we can gain insight into how the filters work by looking at them as being voltage dividers that act differently at various frequencies.

We've just seen that we can use a capacitor and a resistor together to make either a high-pass filter or a low-pass filter, depending on which positions we put them in relative to the ground. An inductor can also be combined with a resistor to make either a low-pass or a high-pass filter. The simplest LR filters are shown in figure 12-8.

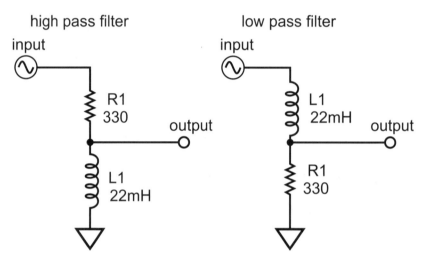

Figure 12-8.

Capacitors and inductors also play a part in impedance. We said earlier that impedance is like resistance: it's the ability to restrict the flow of current, but it differs by being frequency dependent. That variation with frequency comes from capacitance and inductance. The math involved with this relationship is better left for another book. The short version is that any time there is a capacitor involved (or anything that acts like one, by being able to hold a charge), there will be variation in how the circuit reacts at different frequencies. The same is true of inductors, in that if there is anything in the circuit that can store energy in a magnetic field, the circuit will react differently as the frequency changes.

Higher-Order Filters

We've already alluded to the idea that a filter can have a sharper roll-off of response above or below its cutoff frequency—namely, that the filter will have a higher order. We can see this idea in action if we combine the first-order low-pass filters shown earlier. This is shown in figure 12-9. Such a filter will attenuate the unwanted frequencies at −12dB per octave instead of −6dB. The filters shown here are called *second-order filters*.

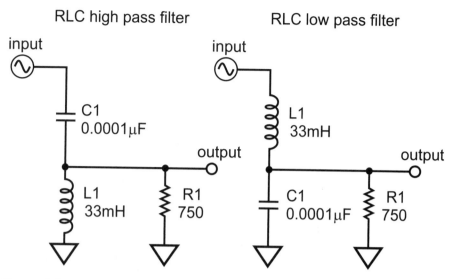

Figure 12-9.

There are other ways of combining capacitors, resistors, and inductors to get even higher-order filters, such as third order, fourth order, and so on. Every engineering solution has its drawbacks, and every filter design involves a price to be paid in terms of phase changes and response smoothness. The details of these choices are not really covered by this book, so we'll move on to looking at a typical use of some filters—namely, the crossover circuit.

Crossovers

Drivers (also called speakers) take an alternating current and turn it into sound. Ideally, a single driver would reproduce every frequency in the human hearing range with equal efficiency. There is not such ideal driver. Some driver designs are good at reproducing bass (the woofer), some at reproducing treble (the tweeter), and some at the range in between (the midrange). Asking the tweeter to reproduce the sound of the low strings on a bass guitar is ineffective, and possibly hazardous to the driver.

Since we want to be able to hear the whole range of sound, we use two or more drivers to make up a full-range speaker. We need to be able to send the treble to the tweeter, the middle frequencies to the midrange, and the bass to the woofer. The simplest way to do this is the *passive crossover*, which is an arrangement of passive (unpowered) components designed to split the overall signal into three portions, one for each speaker element. Inductors and capacitors are the two major components in passive speaker crossovers, where they interact with the impedance of the drivers to create RC filters (resistive-capacitive) and RL filters (resistive-inductive). If we connect the inputs of a low-pass filter and a high-pass filter to the same amplifier, we have a basic passive crossover. The high-pass filter would send the trebles to the tweeter, and the low-pass filter would send the bass to the woofer.

A Typical Passive Crossover

Figure 12-10 shows the circuit diagram of a passive crossover that I extracted from a three-driver home stereo speaker.

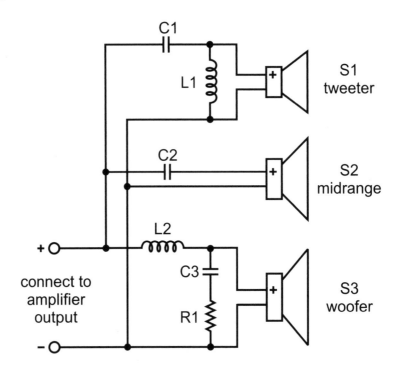

Figure 12-10.

The tweeter is fed from a second-order high-pass filter formed by the capacitor C1, the inductor L1, and the impedance of the tweeter driver S1.

The woofer is fed from a second-order low-pass filter formed by inductor L2, capacitor C3 in series with resistor R1, and the impedance of the woofer driver S3. The resistor R1 has been added to adjust the cutoff frequency.

It would make sense to expect the midrange (S2) to be fed from a combination of a high-pass filter (to remove the lowest frequencies) and a low-pass filter (to remove the highest frequencies). This is how some crossovers are made. In the example crossover, only a first-order high-pass filter is used, with the capacitor C2 in series with the impedance of the midrange driver S2.

An ideal speaker crossover would make a very strict distribution of the signals. For example, everything above 3kHz would be routed to the tweeter, everything from 500Hz to 3kHZ would go to the midrange speaker, and everything below 500Hz would be routed to the woofer. A perfect crossover such as this is not practical, because the filters would have to be of such a high order. Instead, a certain amount of compromise needs to be made, taking into account the frequency response and impedance of each speaker element. The intent is to create a combination of speakers and filters that work together to give a flat response. A graph of a typical tweeter-midrange-woofer crossover output is shown in figure 12-11, showing the crossover points between the three drivers.

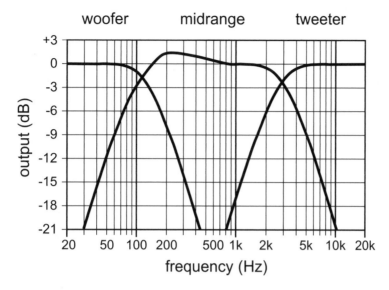

Figure 12-11.

Active Filters

Inductors and capacitors are both passive components, meaning that they exhibit their filtering without needing a voltage supply. It's also possible to build filters from active components, which use transistors and other semiconductors. These *active filters* can act more reliably and accurately than their passive counterparts. A *gyrator* is a combination of active and passive components that can be used to act like an inductor and take its place in a circuit. In many cases the gyrator is the preferred solution to a filtering problem, because real inductors tend to be bulky and imprecise. The intricacies of active filtering, like the complexity of higher-order filters, are beyond the scope of this book. The beginnings of the path to active filters are found in the next chapters, where we start to look at semiconductors.

Simulating How Inductors Work

I would have liked to include a hands-on practice section for this chapter, but it turns out to be impractical. Inductors that are large enough to build up their magnetic fields at a rate that can be observed with a multimeter are too expensive to buy just to use in one practice experiment. If you would like to get some experience with how inductors behave, I suggest using one of the circuit simulators noted in Appendix D. These circuits will demonstrate very admirably how an inductor acts with DC and AC voltages.

13

Diodes and LEDs

Goal: When you have completed this chapter, you will be familiar with how semiconductors work and how these properties are found in the function of a diode.

Objectives

- Differentiate between P-type silicon and N-type silicon.
- Understand the interaction of P-type and N-type conductors in PN junction.
- Recognize the symbol for the diode.
- Build a circuit demonstrating the diode's function as a semiconductor.
- Observe the threshold voltages of diodes and LEDs by measuring their voltage drops.

The Diode Is a Simple Semiconductor

This is the beginning of our journey into active electronics, which allow us to amplify voltage and current. At the heart of these circuits is the *semiconductor*. Here we will learn about the functioning of diodes, which are the most basic semiconductors. This will lead us to the marvels of transistors and amplification in the next chapter.

So far we have worked with materials and components that fall into one of two classes: either they conduct electricity (copper wire, steel, resistors) or they are insulators (plastic, rubber, air). We've also worked with LEDs. These are in a third class, namely the semiconductors. As the name implies, a semiconductor is a substance that only conducts electricity some of the time, under certain conditions.

If you've been doing the hands-on exercises, you'll know from your experience working with light-emitting diodes (LEDs) that they have polarity. They need to be connected in a particular direction in order to function. If we put them in backwards, they don't let any current through. If we put them in forwards, they let current through and also produce some light. This is an example of the conditional nature of a semiconductor because it lets current pass in one direction but not in the other. There are many more complex semiconductors, but the diode is the simplest. It just has one decision to make: on or off.

How Semiconductors Work

We don't really need to know how semiconductors work in order to use them, just as we don't really need to know how the internal combustion engine works in order to drive a car. For years I skipped over this section in the books I was reading, and I got along pretty well. Over time I realized that having a little more knowledge of the physics of what is going on inside the diodes and transistors and IC chips also gave me a better understanding of why they act the way they do, so I started studying it more.

To understand semiconductors, we need to take a brief detour back into the structure of the atom. For this discussion we will be talking about real-world current, the movement of electrons. As shown in figure 13-1, the atom has a central nucleus composed of protons and neutrons. The protons are positively charged. The nucleus is surrounded by a cloud of electrons, which are negatively charged. These electrons are arranged in energy levels, called shells. The first shell can contain 2 electrons, the second shell can contain 8, the third contains an additional 8, the fourth 16. There are more shells, but we don't care about them at the moment, because these larger elements aren't part of the semiconductor story.

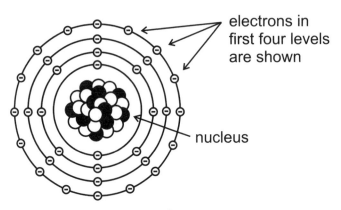

electrons in first four levels are shown

nucleus

Figure 13-1.

It is this arrangement of energy levels that gives us the periodic table shown in figure 13-2. Starting at the top, each row shows the elements arranged by the energy levels of their electrons, with one shell per row. Just as in the diagram of the atom, only the first four energy levels or shells are shown in this table. Note that silicon (Si), boron (B), and phosphorus (P) are highlighted.

1 H									2 He
3 Li	4 Be			5 B	6 C	7 N	8 O	9 F	10 Ne
11 Na	12 Mg			13 Al	14 Si	15 P	16 S	17 Cl	18 Ar
19 K	20 Ca	21-30 Sc-Zn	31 Ga	32 Ge	33 As	34 Se	35 Br	36 Kr

remaining elements not shown

Figure 13-2.

The interactions of atoms is based on every atom having two competing desires. It wants to have its number of electrons balanced with its number of protons, and it wants to have its outermost shell completely filled with electrons. Some elements are naturally in this state, such as helium (He), neon (Ne), argon (Ar), and krypton (Kr). We find these along the right edge of the periodic table. These are called the noble gases, because they don't really react with anything. The reason they don't react is because they are already satisfied. Their electrons match their protons, and their outer shells are filled. They don't need anything else.

All the rest of the atoms have to compromise, always reaching for a state that is closer to the filled outer shell, while still maintaining the electron-to-proton balance. This compromise is the basis of why semiconductors work.

Silicon is the main element in semiconductors. It's one of the most common elements on the planet. We find it in the third row of the periodic table. Counting over from the left side, we see that it has four electrons in its outer shell, and looking to the right, we see that it would like to have eight (like argon). In its pure form, this desire to have eight electrons leads it to form bonds with four other silicon atoms, as shown in figure 13-3.

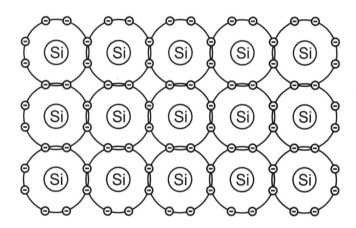

Figure 13-3.

In this structure, it shares its four electrons with its neighbors, and in turn it gets to borrow electrons from them. While it's not as ideal a situation as argon has, this crystal structure gives silicon the best it can get. It also makes it unwilling to let go of any of its electrons to let them move around in the crystal. This makes pure silicon an insulator.

Where things get interesting for silicon is when we add a little bit of something else. For example, let's look at phosphorus. It's right next to silicon on the table, as shown in figure 13-2. It has five electrons in the outer shell. If we mix a tiny bit of phosphorus into the silicon, it gets integrated into the matrix of the crystal, but now the crystal structure has a few extra electrons. Since electrons are negatively charged, we call it *N-type* silicon, where "N" stands for negative.

While the phosphorus atom wants to keep all its electrons with it, when it's in the crystal, it also has four electrons from the nearby silicon atoms. Essentially it ends up with one extra electron that it doesn't want, and when it has a chance, it kicks it out to a nearby silicon. This silicon in turn kicks it out to another silicon when it gets the chance, and so on to the next. In this way, there is movement of electrons from one atom to the next, so current is flowing. The silicon that was previously an insulator is now a conductor. We call it an *N-type conductor*, as shown in figure 13-4.

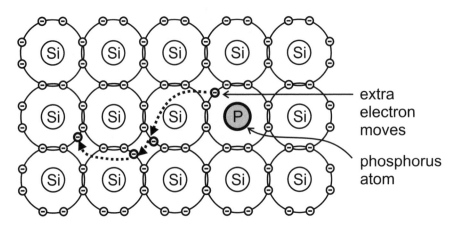

Figure 13-4.

Going back to the pure silicon, we can also add a bit of boron to it. Looking at the periodic table again in figure 13-2, we find that boron is one row up and one column to the left of silicon. Boron has three electrons in its outer orbit. When we mix a little in with the silicon, the crystal now has some spaces in it. This silicon is missing electrons, so it's more positively charged than pure silicon. We call it *P-type silicon*, where "P" stands for positive.

This P-type silicon is just a few electrons short. The boron wants to get ahold of another electron to complete its shell, so it steals one from a nearby silicon atom. In turn, that silicon steals one from another silicon atom. Again, we have the movement of electrons. This is current, so we can call this mixture a *P-type conductor*. This is shown in figure 13-5.

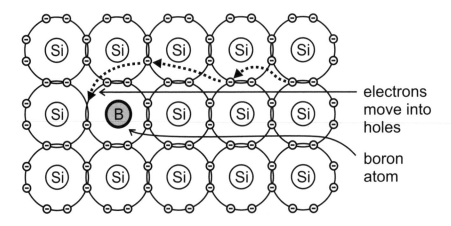

Figure 13-5.

So far, all we've managed to do is go through a very convoluted and expensive process to turn an insulator into a conductor. Individually, N-type and P-type conductors are not particularly thrilling. It's when we put them together that things get interesting. When we put a block of

N-type next to a block of P-type, the excess of electrons in the N-type can fall into the "holes" in the P-type. Current can now flow across the junction from N-type to P-type. However, if we try to have current flow from P-type to N-type, it doesn't work; since the P-type is short on electrons to start with, it doesn't have any extra to donate to the N-type. A block of P-type and a block of N-type is called a *PN junction*, as shown in figure 13-6.

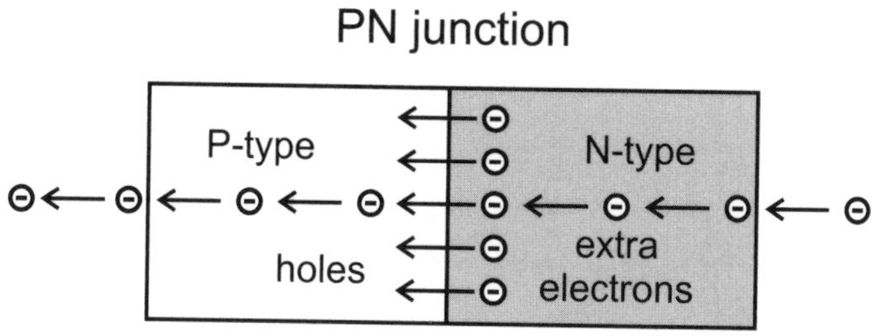

PN junction

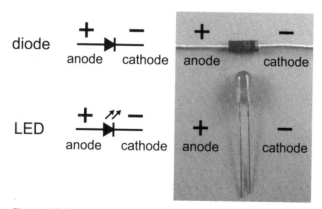

Figure 13-6.

We now return to talking about current as conventional current, which flows from (+) to (−).

The PN Junction in a Diode

A PN junction as just described is the functional part of a diode. The two leads of a diode have specific names: the *anode* is the positive terminal (+) and the *cathode* is the negative terminal (−). The anode should be connected to the incoming current, and the cathode should be connected to the side closest to the signal ground. The industry standard is to mark the cathode end of a diode with a band. (One way to remember which terminal is which is that current flows from anode to cathode, which is *A* to *C*, in alphabetical order.) The symbols for diodes and LEDs with their real-world counterparts are shown in figure 13-7. In a circuit diagram, a diode is usually labeled with a **D**. An LED may be labeled **D** or **LED**.

Figure 13-7.

When the voltage at the anode is positive compared to the voltage at the cathode, we say that the diode is *forward biased*, which results in a flow of current. When the voltage at the anode is negative compared to the voltage at the cathode, the diode is *reverse biased*. In this situation, there is no current flow. It's important to note that the bias, forward or reverse, refers to the polarity of the voltages, not the current.

This is what would happen with an ideal diode. In an ideal diode, maximal current flows to the cathode when any positive voltage at all is applied to the anode, and no current at all ever flows when any negative voltage is applied to the anode. The real world doesn't contain any of these ideal diodes. Diodes have very different types of behavior when presented with various levels of forward and reverse bias.

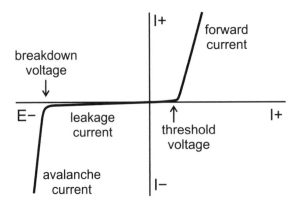

Figure 13-8.

Real-world diodes require a certain level of forward bias (positive voltage) to be able to conduct. When the voltage applied across the diode is below this *threshold voltage*, only a very tiny bit of current passes through the diode. This is shown in the right half of the graph in figure 13-8, which shows current (on the vertical axis) against voltage (on the horizontal axis). Above this voltage, the diode conducts pretty much like any conductor, up to the diode's maximum current rating. Above this rated current, the diode blows up or melts.

When presented with a low level of reverse bias, a real-world diode allows a small amount of current to leak backward through it. This is called the *leakage current*. This is shown in the left half of the graph in figure 13-8. If we continue to gradually increase the amount of reverse bias, at some point the little leak of current will suddenly turn into a flood of current, because the negative voltage of the reverse bias overcomes the semiconductive properties of the PN junction and causes it to become a conductor. This current is called the *avalanche current*, and the voltage at which it occurs is called the *breakdown voltage*. When the diode conducts a limited current in this way, it can recover and return to blocking the current if the reverse voltage is reduced to below the breakdown voltage. If a large current is passed through the diode in this mode, it will explode or melt. (We usually call this "irreversible damage," but that's just a fancy name for destruction.)

The threshold voltage of a diode depends on what kind of material is used to make it and on how it is designed. Silicon diodes, which are the most common, have a threshold of about 0.7V–1.5V. Germanium diodes have a threshold of about 0.3V (300mV). A special design called a Schottky diode has a very low threshold of about 150mV. The threshold of an LED is typically higher, usually 1.5V–2.0V.

A diode or LED doesn't have any ability to limit the current that goes through it. If the current is flowing, the LED will emit light, but running the LED at a very high current will burn it out. If properly protected from high levels of current, an LED will last essentially forever. To protect the LED and provide the right level of current to it, a resistor is placed in series between the LED and the voltage source. This is why we have always included a resistor in series with an LED in the exercises.

The threshold voltage across the LED (or any diode) is seen as a voltage drop across the two terminals of the diode. The result is that the voltage across the resistor is lower than the source voltage. This is shown in figure 13-9. In the exercise on Ohm's law, we saw that we can calculate the resistance needed in the circuit to keep the right amount of current flowing through an LED. In the hands-on exercise for this chapter, you will observe the threshold voltages of LEDs.

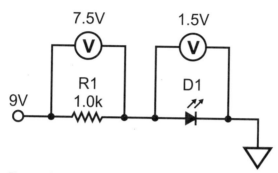

Figure 13-9.

Just like LEDs, diodes (also called rectifier diodes) are subject to burnout from having too much current run through them. A rectifier diode usually has a much higher current rating than an LED. While there are many types of diodes, the most frequently seen is the little black cylinder with the silver band at the cathode end. You will use one of these in the hands-on exercise. This is a 1N4002 diode, which has a maximum current rating of 1.0A in continuous use and a breakdown voltage of 100V.

The diode has many uses in electronics, but the most common is as part of a full-wave rectifier. This is a configuration of four diodes that takes an input of alternating current (AC) and converts it to direct current (DC). We'll explore more about how this works in a later chapter.

Hands-On Practice

The goal of this exercise is to recognize diodes and look at their behavior. You will examine several circuits built with diodes and look at how they behave with varying voltages.

Materials

These are the materials you will need for these exercises.

- Digital multimeter (DMM)
- Breadboard
- 9V battery and battery clip with wires
- 470Ω resistor
- 1.0kΩ resistors
- 1N4002 diodes
- Red LEDs
- Green LEDs
- Insulated solid wire (20, 22, or 24 gauge)
- Optional: grab bag of LEDs

Basic Diode Circuit

In this exercise, you will observe the properties of a diode.

- Use the resistor color code to identify a 470Ω resistor.

- Identify the 1N4002 diode. It's a small black cylinder with two leads and a silver band at one end. The silver band indicates the cathode, the negative end of the diode.

- Construct the circuit shown in figure 13-10.

- There are two components that will prevent the LED from lighting if they are put in backwards: the LED itself and the diode. For the LED, the long lead (the anode) needs to be closer to the voltage supply. The completed circuit is shown in figure 13-11.

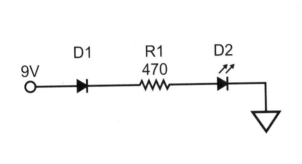

Figure 13-10.

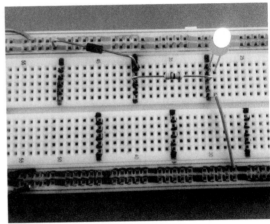

Figure 13-11.

- As you are constructing this circuit, you will find that the diode leads are a little thicker than usually fit comfortably in the breadboard holes. This is normal. Just use a little extra pressure to get the leads in and seated properly.

- Use a piece of wire to complete the circuit from the cathode (the [–] terminal) to the ground bus.

- The LED should light up when the circuit is completed correctly.

- If the LED doesn't light, check that the diode is inserted with the anode connected to the voltage supply and the cathode toward the ground. Also check that the LED is positioned correctly with the anode toward the voltage supply and the cathode connected to the ground.

- Once the circuit is working, measure the voltage across the diode D1. Write down this voltage.

- Measure the voltage across the resistor R1. Write down this voltage.

- Measure the voltage across the LED D2. Write down this voltage.

- Measure the current in the circuit. Remember that you need to open the circuit and complete it with the ammeter to make the current measurement. You should get a current reading of about 13mA. Write down the actual reading.

- With the circuit open, measure the resistance of the resistor R1. Write down the actual resistance you read.

- Use Ohm's law to calculate current, using the actual resistance of R1 and the voltage reading across R1. You should get an answer of about the same value as the current value you read for the circuit.

- Clear the breadboard for the next circuit.

This calculation and measurement bring up an interesting point about diodes and LEDs. These components have a threshold voltage below which they won't let current through at all. The diode or LED essentially "steals" this voltage from the circuit, leaving just whatever is left. If a diode steals 2V and the LED steals 1.8V from the voltage supply, Kirchhoff's voltage law says that all that will be left across the resistor is 5.2V. The resistor doesn't know why the voltage has decreased. It just knows that it is lower, and it restricts current according to Ohm's law. Specifically, it is experiencing a voltage drop of 5.2V, and it's a 470Ω resistor, so the current is about 10mA. What's happening is that the diode and LED are reducing the current in the circuit, not by restricting the current like a resistor, but by reducing the voltage across the resistor.

Measuring Threshold Voltages

This exercise demonstrates the variation in the threshold voltage between diodes and LEDs.

- Use the resistor color code to identify a 1.0kΩ resistor.

- Build the circuit shown in figure 13-12. This is a very simple circuit. Its function is to allow you to measure the threshold voltage of several diodes and LEDs. We'll start with the diodes.

- Measure the voltage across the diode D1. Write down the result. This method is shown in in figure 13-13.

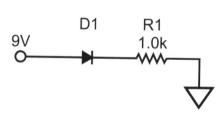

Figure 13-12.

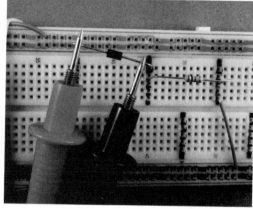

Figure 13-13.

- Remove the diode and replace it with a second diode, D2.
- Measure the voltage across D2. Write down the result.
- Remove the diode and replace it with a third diode, D3.
- Measure the voltage across D3. Write down the result.
- Compare the threshold voltages of the three diodes. They should all be pretty close to one another.
- Remove the diode and replace it with a red LED: D4.
- Measure the voltage across D4. Write down the result.
- Repeat the measurements with two more red LEDs: D5 and D6.
- Compare the threshold voltages of the red LEDs. They should all be close to one another, but different from the other diodes.
- Repeat these threshold voltage measurements on three green LEDs: D7, D8, and D9.
- Compare their threshold voltages to the red LEDs. Again, we expect the green LEDs to be similar to one another but different from the diodes and red LEDs.

Grab Bag of LEDs

If you have a grab bag or other assortment of LEDs, it is interesting to measure their threshold voltages just as you just did with the green and red LEDs.

• These components can vary substantially in their threshold voltages due to their construction, with different metals being mixed in with the silicon to optimize the light output of the LED.

• You can use the same circuit that you used earlier for testing the voltage drops across various LEDs.

• You can expect to see much more variation between the various colors and models of LEDs. In most grab bags, you will find a two-color or three-color LED where there are more than two leads. In this case, either the anode or the cathode is common to all of the LEDs. Some experimentation will be needed to figure out how the leads are arranged.

This set of exercises has given you experience building circuits with diodes and LEDs, including observing their threshold voltages.

14

Transistors and Tubes

Goal: When you have completed this chapter, you will have seen how combining N-type and P-type conductors creates a transistor that allows signal amplification to occur, and how this semiconductor acts similarly to a vacuum tube.

Objectives

- Identify the structure and function of an NPN transistor.
- Recognize the symbol for an NPN transistor.
- Identify the amplification strategies of class A, class B, and class AB amplifiers.
- Recognize the symbol for a vacuum tube.
- Identify the similarities in function of a transistor and a vacuum tube.
- Use a transistor to build a DC amplifier on the breadboard.
- Measure a range of input and output currents of the amplifier.
- Graph this data to identify the linear range and saturation point of the transistor.

Transistors

You have learned about how diodes work, by controlling current flow to be in one direction only. Transistors add to this concept by having one current controlling the flow of another current through the component. We will also explore the types of amplification that these transistors can perform, and the relationship between transistors and vacuum tubes.

The *transistor* is at the heart of the electronic world in which we live. Transistors are everywhere, in every electronic device we use (except for those cool old tube amplifiers, which we will talk about later). The average computer contains about 500 million to 3 billion transistors. A quick glance through any catalog of electronics components will show you that there are thousands of different transistors available, either by themselves or arranged into an *integrated circuit* (IC).

The transistor is the main member of the class of semiconductors. As we saw in the chapter on diodes, semiconductors are not really conductors, but they are not really insulators either. They only allow current to flow under certain conditions. A diode is a semiconductor that only conducts electricity in one direction. A transistor is somewhat like a diode in that it has two contacts between which current can only flow in one direction. Unlike a diode, a transistor also needs current applied to a third contact to let the current flow through the first two. This is shown in figure 14-1. Here we see the symbol for a transistor, showing the flow of current through it. A small current flowing in from the side results in a large current flowing from top to bottom.

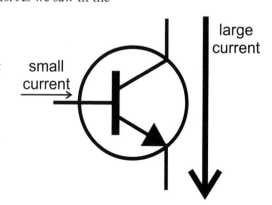

Figure 14-1.

While there are many types of transistors, they all work essentially the same way, allowing a small current to proportionally control a larger flow of current. The key word here is "proportional." There are situations when voltage and current operate separately, but generally an increase in voltage results in an increase in current and vice versa, so we can also say that a transistor allows a small voltage to control a larger voltage. We can envision a transistor as a tiny person sitting on the big water pipe with a big valve on it. The person watches the amount of water flowing from a very small pipe, and adjusts the flow in the big pipe by turning the valve up or down in response to how much water is flowing from the small pipe.

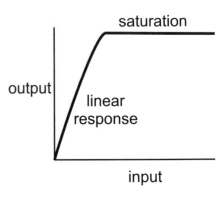

Figure 14-2.

If there is no input current, there is no output current. With a small input current, there is a large output current. With an even larger input current, there is even more output current. This is called a *linear response* or a proportional response because the inputs and outputs form a straight line when they are graphed. Such a graph is shown in figure 14-2. We can keep turning up the input current and keep getting a larger output current until the transistor is at its maximum and just can't give any more output. This point is called *saturation*, and it is responsible for the distinctive sound of distortion we get from an overdriven amplification stage. This is also shown in the graph of figure 14-2. Saturation results in clipping, where the tops and bottoms of a sound wave are clipped off or flattened by not being able to reach their full size.

Why We Need Transistors

In the world of audio, there are a variety of devices that generate very small voltages and currents in response to a stimulus of some kind.

- Microphone capsule, in response to sound.
- Electric guitar pickup, in response to the string vibrating.
- Piezoelectric pickup, in response to the vibration of a guitar soundboard.
- Phonograph stylus cartridge, in response to the movement of the needle in the groove of a record.
- Playback head on an analog tape deck, in response to the magnetic fields on the tape.

All of these things are examples of *transducers*, which are devices that transform one type of energy into another type of energy. The transducers listed here all take physical movement or sound energy as their input and turn it into an output of a small AC voltage. The key word here is "small." Used directly, the output from these transducers is not going to do anything useful, such as driving a speaker or being recorded on tape.

If we provide a transistor with a large voltage supply, the tiny signal from the transducer can regulate how much current is allowed through the transistor. The input signal can be amplified immensely. If the single pass through the transistor doesn't make it big enough, the transistor output can be fed into another transistor and amplified some more.

There are a number of types of transistors, with each type represented by a different schematic symbol. For the moment, you only really need to know about one of them. This is the *NPN transistor*, which is shown the upper corner of figure 14-3. The others shown are examples of other types of transistors, which are included here so that you'll recognize their function if you see them in a circuit diagram. There are other transistor symbols in addition to the ones shown here, but all of them have the distinction of having three terminals where current into or out of one controls the flow between the other two.

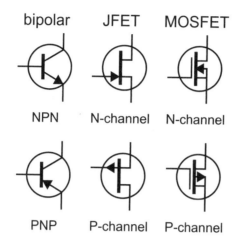

Figure 14-3.

NPN stands for negative-positive-negative, referring to the junction made of a sandwich of two blocks of N-type silicon separated by P-type silicon. These NPN transistors are the most common type of transistor. The next most common is the PNP, which stands for positive-negative-positive. An easy memory aid for distinguishing NPN from PNP is that the arrow of the NPN points out. We can remember this because it is Not Pointing iN. These letters make up the name NPN. These two kinds of transistors are known as *bipolar transistors*.

As the name implies, the NPN transistor is made of a thin layer of P-type silicon sandwiched between two blocks of N-type silicon. This is shown in figure 14-4. Thinking about the PN junction of the diode, we would expect that current flowing into the P-type layer would only flow into both the blocks of N-type, as shown in the second panel. This wouldn't be very interesting, and it also isn't how it works. What actually happens is that if we put a voltage across the two N-types with no current into the P-type layer, there is no conduction across the P-type, so no current flows. If we run a little current into the P-type layer, the P-type become permeable and lets current flow from one block of N-type to the other, as shown in the figure. In this way, a small current (flowing into the P-type terminal) controls a larger flow of current between the other two terminals.

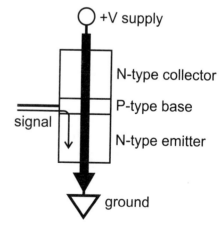

Figure 14-4.

The arrow in the symbol shows in which direction the main current flows through the transistor and where the current flowing into the middle of the sandwich goes. Like the diode, the current only flows in one direction. NPN and PNP transistors have three contacts. They are named the *base*, the *emitter*, and the *collector*. In these bipolar transistors, the collector and the emitter are like the big pipeline and the base is like the little controlling pipeline. In other types of transistors, the three contacts have different names but essentially the same functions.

Some transistors are shown in figure 14-5. At the left are small-signal transistors. The amount of heat they generate is low, though caution should still be taken not to crowd them together or restrict the air flow around them. At the lower center is a larger transistor with a built-in heat sink. At the upper center and the right are even larger transistors that must be mounted on a piece of metal to radiate and dissipate the excess heat produced by the transistor. The pins of the transistor on the right are hidden underneath. In a circuit diagram, a transistor is typically labeled **Q.** We'll see this in the circuit in the hands-on exercise. In some circuits they are labeled **U.**

Figure 14-5.

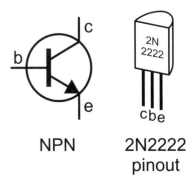

NPN   2N2222 pinout

Figure 14-6.

The diagram in figure 14-6 shows the pin configuration (called the *pinout*) for a particular-example NPN transistor. This is the 2N2222 transistor you will be using in the hands-on exercises.

There is no standardization in the transistor industry as to which pin is connected to the base, emitter, and collector of the NPN silicon sandwich. To determine the pinout of any particular transistor or any other semiconductor, one should refer to the *datasheet*, also called the *specification sheet* or just "the specs." Thanks to the Internet, datasheets are much more easily available than they used to be. Many manufacturers and distributors have them available on their websites. The datasheet also shows operating conditions, maximum voltage, and all the other things you might need to know about a component to understand how to use it in a particular circuit design. Incidentally, don't let the volume of information on a datasheet over-whelm you. They're showing you all of the information that electrical engineers might possibly ever need to know for using the components in a particular design. Usually you just need one or two of those pieces of information, such as which pin is which. As you work more with the components, you'll find other specifications to be useful, but at the moment, the majority of the information is not relevant to the work that you are doing.

Amplifier Classes

We can use a transistor to let a small current control a large current, and that can be used for amplification. Here we will look at how this works and how this is reflected in the classification of amplifiers.

An AC signal is a voltage that varies above and below the ground: +1V, 0V, –1V, 0V, and so on. Stated another way, the AC signal is made up of positive, zero, and negative voltages. If we have a load attached to the voltage, there's a current of matching polarity that flows through that load: +10mA, 0mA, –10mA, 0mA, and so on. The current flowing through the load is either positive, zero, or negative.

First we'll look at the three types of inputs we can feed into an NPN transistor. If we give an NPN transistor an input of a small positive current at its input, it will give us a large positive output. This is shown in figure 14-7. If we give an NPN transistor no input (zero current), it will give us no output (zero current). This is shown in figure 14-8.

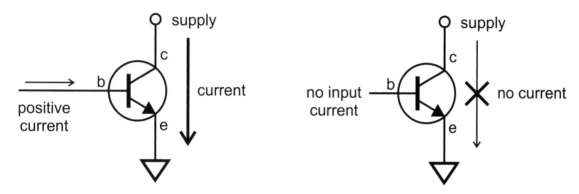

Figure 14-7.   Figure 14-8.

Our audio signal is made up of positive and negative voltages, resulting in positive and negative currents. If we give the NPN transistor a small negative input current (or a large one), we would like it to give us a large negative output. Unfortunately, it's not going to do that, as shown in figure 14-9. Remember that an NPN transistor is like a diode: current only flows through it in one direction, from top to bottom. So if we give an NPN transistor a negative input current, it will give us no output (zero current).

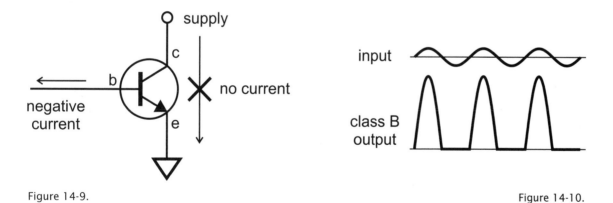

Figure 14-9.   Figure 14-10.

Clearly, only amplifying the positive half of the wave is not going to work for accurately amplifying an audio signal. Amplifying just half of the wave is called *class B amplification*. An example is shown in figure 14-10. This type of amplification does make the signal larger, but in the process it also deletes the bottom half of the wave. This is clearly unacceptable for working with an audio signal.

Class A Amplification

One solution here is to add a certain amount of positive DC voltage to the input signal. This voltage is called the *bias voltage*, as shown in figure 14-11.

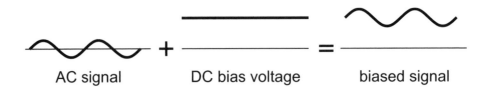

AC signal + DC bias voltage = biased signal

Figure 14-11.

When we use this biased signal as the input for our transistor, we get accurate amplification of the signal, because all of the input current is positive. This is called *class A amplification*, which is defined as a single transistor (or tube) providing all of the amplification for that gain stage. The result of using this biased input is shown in figure 14-12. You've probably seen preamps in the studio that say "100% Class A amplification" on them. This means that the entire signal path uses this type of amplification.

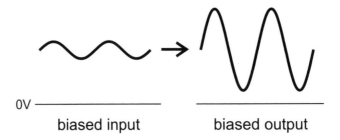

0V —— biased input   biased output

Figure 14-12.

Now we have a different problem, because while the output is larger, it is also biased. The amplifier output contains the AC signal we want, but it also contains the DC offset. Fortunately this problem is fairly easy to fix. Remember that one of the functions of a capacitor is passing AC while blocking DC, which is precisely what we need here. We want the AC part without the DC part, so we will use a capacitor in series with the output to filter out the DC, as shown in figure 14-13. Typically the input of a gain stage will have a capacitor in front of it to remove any bias that may be part of the signal. This is called a decoupling capacitor.

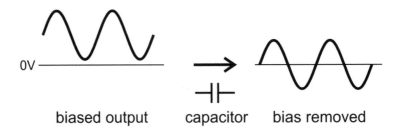

0V —— biased output   capacitor   bias removed

Figure 14-13.

Class AB Amplification and Single-Ended Amplification

A different amplification strategy is to use two class B amplifiers together. One of them amplifies the top half of the wave, and the other amplifies the bottom half. When these two amplified signals are added together correctly, the combined signal is an accurate amplification of the original input signal. This is called *class AB amplification*, because it uses two class B amps acting like a class A. This is shown in figure 14-14.

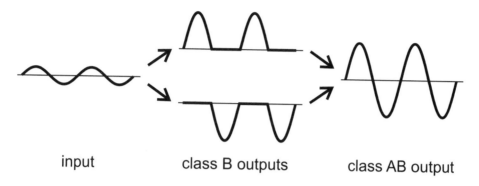

input   class B outputs   class AB output

Figure 14-14.

If class AB is so complicated and finicky, we might ask why we don't we just use class A for everything, such as power amps, preamps, guitar amps, and so on. At first it seems obvious that all amplification ought to be class A, because there is so much less to go wrong with it. We need to consider what makes class A possible—namely, the bias voltage.

Whatever bias voltage we use will be proportional to the output needed. If we set it too low, the signal will clip at the bottom. The consequence is that for high voltage output, the bias voltage also needs to be quite high. In turn, even when there is no input at all, there is still a considerable current flowing through the transistor. In a mic preamp this current will be fairly small, but in a power amp it will be substantial, so that a lot of heat is generated and power is lost from an amplifier that is just sitting there doing "nothing." This means that almost all power amps are class AB or some even more esoteric configuration.

With that said, there are a few power amps that use class A. These are usually called single-ended power amps, and a lot of them are tube amps. There are a few audiophiles who are dedicated to this design, even though it's so inefficient. These single-ended power amps are not typically found in the studio environment.

Vacuum Tubes

Before there were transistors, there were tubes. Formally, these are called *vacuum tubes*. The British name for a tube is a "valve." Tubes were the first devices that were used to create amplification. In many ways they are quite different from transistors, but both devices perform the same tasks of selectively managing the flow of current. Semiconductor diodes and tube diodes were developed at about the same time in the 1870s, but tubes became more popular because they gave more consistent results. An example of a vacuum tube is shown in Figure 14-15.

To understand how a tube works, first we need to specify that for this discussion we will be talking about real-world current flow, which consists of the electrons from the negative of the voltage source flowing to the positive. We start with two metal plates with a small air gap in between. If we hook up a voltage supply of 300V across them,

Figure 14-15.

we now have a group of electrons on the negative plate (the cathode) that desperately want to get to the positive plate (the anode). They can't get there, because of the air gap. It turns out that there are several ways in which we can help them get across, and the tube is the embodiment of these measures.

We can put a heating filament right next to the cathode and heat it up to white-hot. This causes the electrons to stream off the cathode toward the anode. Unfortunately, they run into the air molecules in the air gap. The solution is to enclose the anode, the cathode, and the filament in a glass tube and remove all of the air. This makes it a vacuum tube, as shown in figure 14-16.

We now have the original diode. The name is from the Greek *di* for "two" and *ode* for "path." Electrons flow from the cathode to the anode. In the days before transistors, this tube was used for changing AC supply voltage to DC supply voltage, as will be described later in the chapter on rectification. Two of the symbols for this tube are shown in figure 14-17.

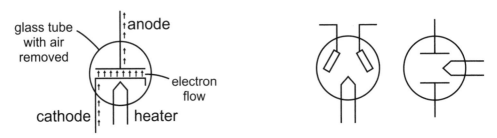

Figure 14-16.   Figure 14-17.

This diode tube does not yet meet the description given earlier for the function of a tube, which is to manage the flow of current by letting a small voltage control a larger current. To do that, we need to make some modifications to the diode tube. First we'll move the anode and the cathode apart a little bit, reducing the current flow from cathode to anode to essentially nothing. Keep in mind that we still have a group of very eager electrons on the cathode that want to jump across the gap to the anode. We can think of these electrons as trying to jump off of the cathode to get to the anode, but they are falling back down because we've widened the gap too far. This is shown in figure 14-18.

We can place a wire grid or screen in between the cathode and the anode. The formal name for this part is the *control grid*, but it often just gets called the grid. This third path makes the tube a *triode*, meaning literally "three paths." The symbol for the triode is shown in figure 14-19.

Figure 14-18.   Figure 14-19.

We can place a small voltage on this grid, perhaps one-tenth of a volt. This low voltage gives a few of the eager electrons that are already being pulled toward the anode a little extra pull, and they can escape the cathode and head for the grid. Having gotten that far, they fly through the grid and go to the anode. Now we have current flowing from cathode to anode. This is shown in figure 14-20.

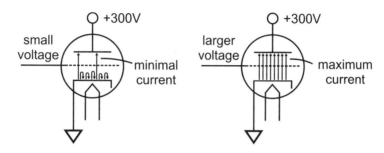

Figure 14-20.

If we increase the voltage on the grid, more of the electrons will be able to escape and make the trip to the anode. Similarly if we turn down the voltage, the flow of electrons will decrease. This is exactly the description we were looking for—namely, of a small voltage controlling a large current. There is a limit to how many electrons can flow across the gap, just as there is a current limit in a transistor. When a tube reaches the point of not being able to respond by letting through more current regardless of an increase in input voltage to the grid, we say that the tube is in a state of *saturation*. It is this state of saturation and how the tube acts as it approaches and reaches this point that give it the unique "tube sound" that audio producers, engineers, and guitarists find so appealing.

The triode is the most basic tube configuration for amplifying a signal or acting as a switch. The engineers who designed these tubes found that additional grids improved their performance in some situations. Adding one more grid, called the screen, made the tube a *tetrode* (literally "four paths"). Adding yet one more grid, the suppressor, gives us a *pentode* (literally "five paths"). The complexities and functions of these grids are beyond the scope of this book. It is sufficient to say that triodes, tetrodes, and pentodes all do the same thing, which is allow a small voltage on the grid to control a large current between anode and cathode. The symbols for these three types of tubes are shown in figure 14-21.

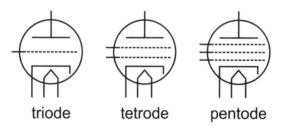

triode   tetrode   pentode

Figure 14-21.

Some tubes have just one set of anode, cathode, grid or grids, and heater. An example is the EL34 tube, frequently used for power amplification. The EL34 is classified as a power pentode. The pinout of the EL34 is shown in figure 14-22, along with the diagram of the octal socket that is used for this tube. Octal refers to the eight pins. Many tube pinouts label the anode with an **a** and the cathode with a **k** for *katode*, which is "cathode" in German. The heater may be labeled **h** for heater or **f** for filament. Grids are typically labeled **g1, g2,** and **g3.** Additionally we see that the circle representing the glass tube can also be shown as an oblong.

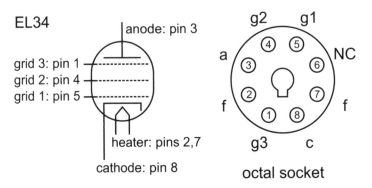

Figure 14-22.

Other tubes combine two sets of anodes, cathodes, and grids into one tube, with a common heater for both cathodes. An example is the classic 12AX7 tube, which is classified as a dual triode. Its pinout is shown in figure 14-23. In schematics, the two halves of the tube may be separated for convenience of showing where they are in the circuit. The heater filament connections are often not shown along with the anode, cathode, and grid. Instead they may be shown closer to their connections to the voltage supply. This can be a little confusing, because it means that one physical object, the tube, is shown in three separate places on the schematic diagram.

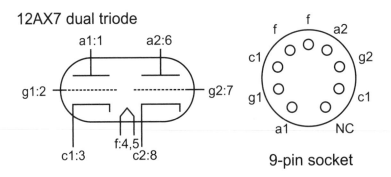

Figure 14-23.

I would be remiss not to mention again the dangers of working on tube amplifiers. Most tube amplifiers use very high voltage supplies, on the order of 300V. These supplies are stabilized and smoothed with big capacitors, as will be seen in the chapter on transformers and rectification. Even with the unit unplugged, the capacitors will take a while to release their charge. Never just assume that a capacitor in a tube unit is safe to touch. The size of the capacitor combined with the large voltage can yield a lethal combination. If you are working inside a tube unit, get in the habit of routinely checking the voltage across the big power supply capacitors before you start doing anything else inside. This habit could save your life. I don't want to scare you out of not working on this equipment, but I do want you to work safely and develop habits that will keep you working safely.

The "Starved Plate" Configuration

Some tube amplification designs use a lower voltage supply instead of high voltage. These designs are classified as the "starved plate" configuration. In this design, the voltage supply requirements are much lower, and the heater supply is also much lower, so the whole circuit can be run off of a little "wall-wart" adapter instead of a big, expensive dedicated power supply. This makes the gear more affordable, but not necessarily better. It may make me a tube-gear snob, but I don't think it's a real tube amplifier unless there's high voltage and the heater is glowing with that classic orange color. In a blatant attempt to overcome this shortcoming, one company included a little red LED behind the tube to provide the illusion of it being a piece of real tube gear. There are vent holes in the chassis so the user can see the tube and be suitably impressed. Unfortunately, real tubes glow orange instead of red, so the illusion is somewhat spoiled.

How Far We've Come from Tubes to Transistors

Just to illustrate how much the modern transistor has changed our electronic world, let's consider the difference between the old and the new. The first real computer was called ENIAC, and it had 17,468 tubes. It was designed during World War II and built under contract with the US military. It was completed just at the end of the war. This elementary computer was used for solving complex math problems, mostly to do with building nuclear bombs and calculating missile trajectories. The iPhone 4 has about 42 million transistors. If we were to build an iPhone 4 with tubes, it would take up a three-story building covering a full city block, and it would consume about 190 megawatts, which is one-sixth the output of the average nuclear reactor. Based on the initial cost of the ENIAC computer, the cost of such a device would be somewhere around $12 billion.

Hands-On Practice

The goal of this exercise is to build a circuit that demonstrates the transistor's ability to amplify current. We'll do this by building a circuit and testing its response with several current levels. This level current will be controlled using different resistors.

Materials

These are the materials you will need for these exercises.

- Digital multimeter (DMM)
- Breadboard
- 9V battery and battery clip with wires
- 470Ω resistor
- 1.0kΩ resistor
- 10kΩ resistor
- 22kΩ resistor
- 47kΩ resistor
- 100kΩ resistor

- 220kΩ resistor
- 470kΩ resistor
- 2N2222 NPN transistor
- Red LED
- Green LED
- Insulated solid wire (20, 22, or 24 gauge)

DC Amplifier Procedure

The point of this exercise is to demonstrate that for a very small input current, the transistor gives a larger and proportional output. At some point it reaches a limit of how much current it can give, and it doesn't give any more regardless of the input current. Your goal is to observe the range of inputs that give a proportional output and compare the levels of the input current and output current. This is called a DC amplifier because it is amplifying the amount of DC current.

- Before starting, test your red and green LEDs with a 1.0kΩ resistor to make sure that they both light up, just as you did in the previous hands-on exercise with LEDs.

- The schematic for this DC amplifier circuit is shown in figure 14-24. In this diagram, R2 is shown as being variable. Later diagrams will show specifically what value to use for this resistor as the experiment progresses.

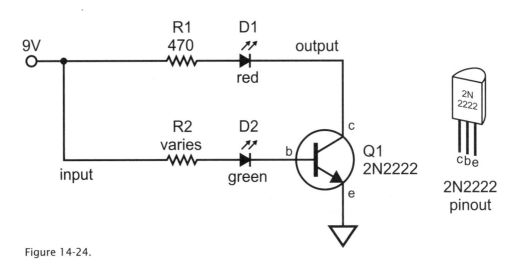

Figure 14-24.

- Up to now, our strategy for building the circuit has been to start with the voltage supply. That strategy is always valid, but in this case it is easier to start with the transistor. The pinout is shown in figure 14-24. Place the 2N2222 transistor (designated Q1) in the breadboard with the base (the center lead) in a marked line in the lower center of the breadboard. The flat side of the transistor case should be facing away from you. We can think of the circuit as being composed of three arms, one each to the three leads of the transistor.

- Place the red LED (D1) into two of the lower sets of component strips, a few columns to the right of the transistor, with the cathode (the short lead) facing toward the transistor. Use a piece of wire to connect the cathode (the short lead) to the collector of Q1 (the lead on the right). It's more convenient for the later connections to run this arm of the circuit along the lower half of the breadboard, off to the right. This is shown in figure 14-25.

Figure 14-25.

- Use the resistor color code to identify a 470Ω resistor, which is R1.

- Connect R1 to the anode of the red LED (D1). Place the other end of R1 in a component strip in the upper half of the breadboard. This is shown in figure 14-26.

Figure 14-26.

- Use a piece of wire to connect the free end of R1 to the voltage supply bus. This completes the first arm of the circuit, as shown in figure 14-27. This is the "output" arm of the circuit, where the current controlled by the transistor flows. You do not expect the LED to light up at this time. Later you will be unplugging this wire to measure the level of the output current.

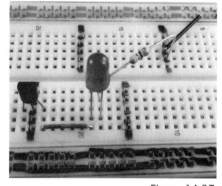

Figure 14-27.

- Insert the green LED (D2) into two of the component strips in the upper part of the breadboard, above the transistor Q1, with the anode to the right and the cathode to the left.

- Use a piece of wire to connect the cathode (the short lead) of the green LED D2 to the base lead of the transistor Q1. Remember that the base lead is the one in the middle. You do not expect the LED to light up at this time.

- Use the resistor color code to identify a 1.0kΩ resistor.

- The schematic shows the value of resistor R2 as being variable. For the first version of this circuit, you will use the 1.0kΩ resistor for R2.

- Connect this resistor R2 to the anode (the long side) of the green LED D2, and plug the other end into any convenient unused component strip on the upper half of the breadboard, to the left side of D2.

- Use a piece of wire to connect the free end of R2 to the voltage supply. This completes the second arm of the circuit, as shown in figure 14-28. This is the "input" arm of the circuit. The small current that controls the larger current through the transistor flows through this arm made up of R2 and D2. (No, *Star Wars* fans, I did not plan that, it's just the way it worked out.) You

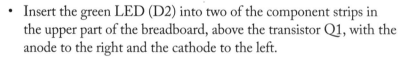

haven't connected the third arm yet, so you don't expect either LED to light up. Later you will be unplugging this wire to measure the level of input current.

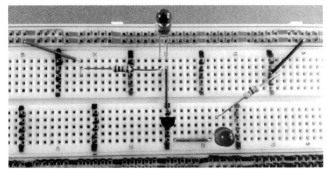

Figure 14-28.

- Use a piece of wire to connect the emitter of Q1 (the lead on the right) to the ground bus. This completes the third arm of the circuit. This is shown in figure 14-29. You expect to see the red LED glowing, and the green LED will be dimly lit.

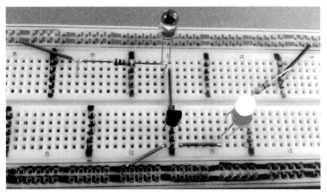

Figure 14-29.

- If you don't see a glow from the LEDs, check that both of your LEDs are the right way around, with their short leads and their flattened sides connected to the transistor Q1. Also check that your 9V battery is providing adequate voltage (8.5V or more) by using your DMM to measure from the voltage supply bus to the ground bus. Also check that the flat side of the transistor Q1 is facing away from you, toward the voltage bus. If none of these things help, check that all of the nodes of the circuit are connected, starting with the voltage supply and going all the way down to the ground. All of the connections down in the body of the breadboard will be between leads plugged in vertically, not next to each other.

- With your circuit working, it's time to start measuring current and recording your measurements. You will be collecting 16 points of data and using them to draw a small graph. You can collect the data in a table and draw it on regular paper following the format of the data table shown in table 14-1 and the blank graph shown in figure 14-35. If you prefer, you can print out the blank data table and blank graph from the supplemental materials on the DVD.

input resistor R2 (W)	input current in mA (green LED D2)	output current in mA (red LED D2)
1.0k		
10k		
22k		
47k		
75k		
100k		
220k		
470k		

← do not graph this point

Table 14-1.

- The first reading is the input current. The position of the ammeter is shown in figure 14-30.

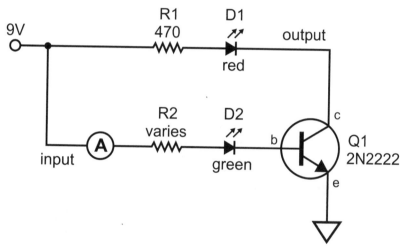

Figure 14-30.

- Set your DMM on the 20mA scale.
- Remove the wire connecting the R2-D2 input arm of the circuit (with the green LED).
- Measure the current in this arm of the circuit and record the reading in the first space in the table. Both LEDs should turn on when you connect the meter to take the reading. The current should be about 6.5mA. This measurement is shown in figure 14-31.

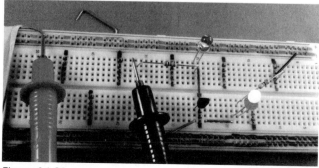

Figure 14-31.

- Remove the meter and reconnect the wire to restore the circuit.
- The next reading is the output current. The position of the ammeter is shown in figure 14-32.

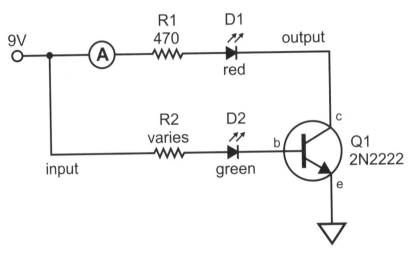

Figure 14-32.

- Set your DMM to the 20mA scale.
- Remove the wire connecting the R1-D1 output arm of the circuit (with the green LED).
- Measure the current in this arm of the circuit and record the reading in the first space in the second column in the table. Both LEDs should turn on when you connect the meter to take the reading. The current should be about 16mA. This measurement is shown in figure 14-33.

Figure 14-33.

- Remove the meter and reconnect the wire to restore the circuit.

You have completed the first set of readings. The next part of the exercise is to repeat the measurements of the input current and the matching output current for each of the R2 resistors. The steps described here will be the same for each resistor.

- Identify the next resistor. In this case it's the 10kΩ resistor.
- Remove the present R2 input resistor, and insert the next resistor in its place. For the 10kΩ resistor, the input LED will not light, and the output LED will be dimmer. Later, when the value of R2 is higher, the output LED will not light either, but you will still be able to detect current by measuring with the ammeter.

- The next reading is the input current. The position of the ammeter was shown earlier in figure 14-30.
- Set your DMM on the 2mA scale.
- Remove the wire connecting the R2-D2 input arm of the circuit (with the green LED).
- Measure the current in this arm of the circuit and record the reading in the appropriate space in the table. If the output LED was lit before, it will light again. This measurement was shown in figure 14-32.
- Remove the meter and reconnect the wire to restore the circuit.
- The next reading is the output current. The position of the ammeter was shown in figure 14-33.
- Set your DMM to the 20mA scale.
- Remove the wire connecting the R1-D1 output arm of the circuit (with the green LED).
- Measure the current in this arm of the circuit and record the reading in the second space in the second column in the table. If the output LED was lit before, it will light again. When R2 is 10kΩ, the current should be about 14mA. For the other values of R2, you expect to see lower current readings, down to less than 3mA for the final R2 resistor, 470kΩ.
- Remove the meter and reconnect the wire to restore the circuit.
- Repeat this set of steps for each resistor shown in the table: 22kΩ, 47kΩ, 75kΩ, 100kΩ, 220kΩ, and finally 470kΩ. As the resistance of R2 increases, the current entering the base of transistor Q1 decreases, because of Ohm's law, which tells us that more resistance means less current. Consequently we will also see less current being allowed through the transistor.
- Plot these data points for the currents on the blank graph, as shown in figure 14-34. You should get a graph similar to the one shown at the end of the exercise. Do not include the first data point, for 1.0kΩ, on the graph. If you are not familiar with graphing, follow the instructions in the next step.

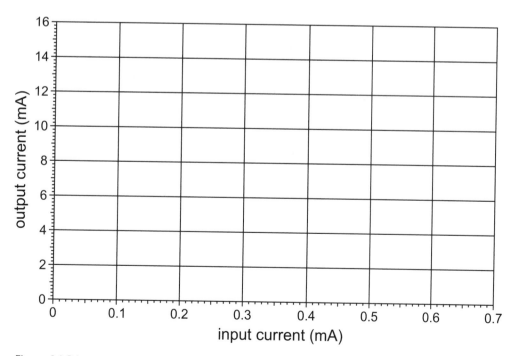

Figure 14-34.

• Each pair of input and output currents you collected in the table is represented by a single data point. Do not try to graph the first data point, with the input and output current for 1.0kΩ. There is no place on the graph for this first point. Start with the input and output currents for 10kΩ. Find the position of the input current for that point on the bottom scale, and draw a very faint line in pencil up at that position. It helps to have a ruler or straight edge for this step. Find the position of the output current for the same point on the side scale, and draw a faint line across until it meets the vertical line. The point where the two lines meet is your data point, so draw a dot there. Erase the faint lines. When you have plotted all the points in this way, connect the dots with straight lines or a curve.

• This graph shows some interesting features of the transistor. One observation is that there is a range of input currents where the output is proportional. For every increase in the input current, the output is increased also. We see this range at the low end of the input currents, where every small increase in input gives a huge increase in output. This is called the *linear region* or *linear range* of the transistor.

• Another interesting feature is what happens when the transistor no longer responds proportionally. At the top of the graph, we see the output current level out, so that the output stays essentially the same no matter how much input current is used. This maximum is called *saturation*. This term is also used for the behavior of a tube with maximum current flow.

This exercise has shown the behavior of a transistor, with its ability to increase or decrease its output current based on its level of input current. The next exercise with a transistor will demonstrate how this variation of current can be translated into variation in voltage, resulting in the amplification of a small signal.

Example

Table 14-2 shows an example of DC amplifier current data.

input resistor R2 (W)	input current in mA (green LED D2)	output current in mA (red LED D2)	
1.0k	6.15	14.75	← do not graph this point
10k	0.63	14.54	
22k	0.286	14.47	
47k	0.135	14.39	
75k	0.081	14.26	
100k	0.064	12.91	
220k	0.03	6.27	
470k	0.014	2.93	

Table 14-2.

Figure 14-35 shows an example of the graph created from the data shown in table 14-2.

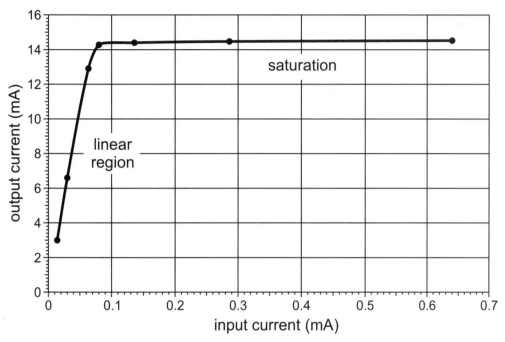

Figure 14-35.

15

Transformers and Rectified AC

Goal: When you have completed this chapter, you will be familiar with the functions of the transformer, the rectifier, and the capacitor in the DC voltage supply.

Objectives

- Identify the function of a transformer.
- Recognize the symbol for the transformer.
- Identify the function of a voltage regulator.
- Test and modify a transformer for use as an AC source for later circuits.
- Build a circuit for the observation of a diode used as a half rectifier.
- Observe the effect of adding a capacitor to this circuit to remove ripple.
- Build a circuit for the observation of four diodes used as a full-wave rectifier.
- Observe the effect of adding a capacitor to this circuit to remove ripple.

Getting from AC Supply Line to DC Supply

A configuration of components dedicated to changing high-voltage AC to low-voltage DC is very common in audio electronics. All the electronic gear you use in the audio world runs on low-voltage DC, such as +9V or +12V. What you get from the wall outlet is high-voltage AC at 120V. These don't match up at all. We use a combination of components to get the high-voltage AC down to low-voltage DC to run a preamp or a console. These components are a transformer, a full-wave rectifier, a capacitor, and a voltage regulator. Collectively these are called the power supply or simply the *DC supply*. In schematics of pro audio gear, they are often called the *power supply unit*, which is shortened to *PSU* for convenience.

The Transformer

The first component in a power supply is the *transformer*. Figure 15-1 shows the symbol for an iron-core transformer. If it reminds you of the symbol for the inductor, that's to be expected. It works on the same principle.

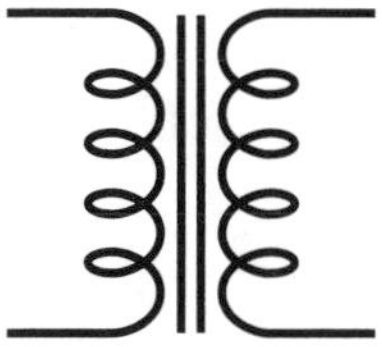

Figure 15-1.

This component is a combination of two (or more) coils of wire both wrapped around a bar or ring of iron. The supply line current from the 120V outlet runs through one of these coils. This coil, which is attached to the input, is called the *primary*. When alternating current (AC) runs through the primary, it generates a magnetic field in the iron core, storing its energy there. This magnetic field generates a current in the output coil, which is called the *secondary*.

At first glance this seems like a useless device; we've changed the current to a magnetic field and back to a current. At best, it seems wasteful. The trick is in the number of turns of wire in each coil. The voltage generated by a magnetic field is proportional to the number of turns in the coil. If we have only a quarter as many turns in the secondary coil as in the primary coil, the secondary can only generate a quarter of the voltage that is applied to the primary.

Therefore, if we connect 120VAC to a transformer with four times as many turns on the primary side as the secondary side, the output will be one-fourth of the input voltage, namely 30VAC. This is called a *step-down transformer*, because it steps down the voltage.

What if we turn the transformer around and connect our 120VAC supply to the side with just a few turns? Logically, we would get a four-fold increase in voltage at the output. This turns out to be the case, with an output voltage of 480VAC. This is called a *step-up transformer*.

We should be able to attach a little 10V AC generator to a 1:12 transformer so we can get 120VAC to power a toaster or a refrigerator. We can, but it doesn't do us very much good. The laws of physics don't let us have anything for free. What we gain in voltage we lose in current. To get 1A at 120V, we would have to provide 12A at 10V, plus a little bit, because there is always a little bit of loss when we change the electricity to magnetism and back again. That's a lot of current, and we'd have to work very hard (with several people on exercise bikes with generators) to provide enough current to the transformer to run a toaster. We can change AC that has high

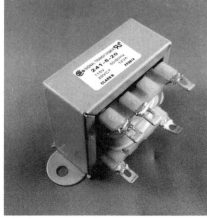

voltage with low current to AC that has low voltage with high current, and vice versa, but we can never put in low voltage with low current and get back higher voltage with higher current. Incidentally, this is why it's hard to use solar panels effectively to provide power for a house or office building. Solar panels produce low-voltage, low-current DC, and we need high-voltage AC.

A typical transformer is shown in figure 15-2. The transformer will usually be labeled as to what input voltage it expects and what output it will provide when given that input.

We recognize transformers by their having four or more wires and a rectangular or donut-shaped winding. In a circuit diagram, a transformer is labeled with a **T** or an **X**.

Figure 15-2

The Center-Tapped Transformer

The symbol shows the simplest form of transformer, with one primary and one secondary. However, if we wanted two stepped-down voltages, there is no reason not to add a second coil on the output side. This would give us two independent secondary coils. This is shown in figure 15-3. These secondaries can either be used separately or in conjunction with each other. A typical way of wiring two secondaries is to tie two of the wires together to form a central tap, as shown in the figure.

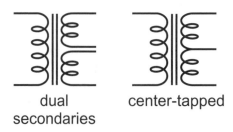

dual
secondaries

center-tapped

Figure 15-3.

Figure 15-4 shows the label of the transformer from figure 15-2. It indicates that the transformer is designed for a 115V primary, as shown on the line just below the model number. The secondary is listed as 20VCT on the next line, indicating that the secondary is 20V and center-tapped.

Now we have the two secondary coils in series, each capable of generating a certain voltage. This arrangement provides a number of possibilities for providing low-voltage supply to a piece of equipment. Figure 15-5 shows that that if the output of each small coil is 20VAC, we can provide either two 10V supplies or a single 20V supply.

Figure 15-4.

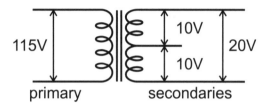

Figure 15-5.

When a Transformer Is More Than Just a Transformer

We also find transformers in the ever-present "wall-wart" converters that are the voltage source for some equipment, such as that shown in figure 15-6. The faceplates of some of these transformers are shown in figure 15-7. Notice that the output on some of these transformers is listed as AC, and others are listed as DC. For any particular piece of gear, both the voltage level and the voltage type (AC or DC) need to be correct.

Figure 15-6.

Figure 15-7.

In the marketplace, two different pieces of equipment are called transformers. The first is the transformer described earlier with the copper wire wrapped around the iron core, all by itself. The second is the whole combined package of transformer, rectifier, and capacitor. Be careful that you know which one you are looking at. If the output is DC, the device that gets called a transformer is actually the package of all three components. These units are also called power supplies and AC adaptors. There are two other useful pieces of information on these faceplates. The first is the current limit on the output. We see 200mA, 700mA, and 1300mA. This is the maximum current that can be provided by the power supply. Another piece of critical information on a DC supply is the polarity of the plug. On the 10.5V Ryobi supply, the tip of the plug is the (+) and the sleeve on the outside is the (–). On the 4.5V Marantz supply, the little drawing in the lower-left corner shows that the tip of the plug is the (–) and the sleeve on the outside is the (+).

Rectification

A transformer provides lower voltage, but it's still AC, and we need DC voltage to run audio gear. The next step is to change the AC voltage to DC. To do this, we use a device called a *rectifier*. The word *rectify* means to fix, to adjust back into alignment, or to correct. We can think of the rectifier as correcting the AC voltage into the proper DC voltage we need for audio electronics.

What we are starting with is AC voltage, which rises and falls above and below the ground. What we want is DC voltage, which is always the same level above or below the ground. For this discussion, we'll focus on producing a clean positive voltage from the constantly changing AC voltage. The rectifier doesn't get us all the way to this goal, but it gets us partway there by eliminating the negative voltage. Overall, the voltage is positive.

The simplest rectifier is a diode. As you will remember, the job of the diode is to allow current to flow in one direction but not in the other. This is perfect for the job of rectifying the AC voltage, because we want only the positive half of the AC wave. If we use a single diode with an AC input, we will get a very bumpy-looking output, as shown in figure 15-8. This output is technically DC, since its average is positive and never falls below the level of 0V at the ground.

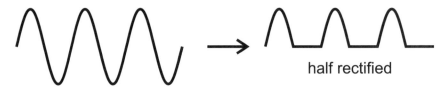

half rectified

Figure 15-8.

This treatment with a single diode is called *half rectification*, because only half of the input wave voltage is used. While the output is the desired DC voltage, this is an inelegant and inefficient solution. We've gone to a lot of trouble to generate this AC voltage. We've built coal-fired power plants, hydroelectric dams, nuclear power plants and a massive distribution system that reaches almost every structure in the continent. With half rectification, when the voltage is finally delivered to us, we ungratefully ignore half of it and throw it away. Clearly this is not a very efficient solution to the problem.

Full-Wave Rectification

The solution is to find a way to harvest the other half of the wave, to flip it from negative to positive. This *full-wave rectifier* is made up of four diodes in a diamond shape, with all of the diodes pointed toward the top. As shown in figure 15-9, this point is the (+) DC out. The bottom of the diamond is the (–) DC out, which we will establish as the ground if we want a positive supply voltage. The AC inputs are at the sides.

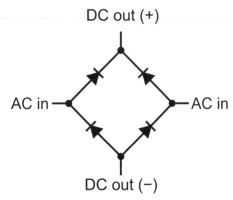

Figure 15-9.

The voltages at the AC inputs are constantly fluctuating up and down. Since neither input is tied to the ground, as one goes up, the other goes down relative to it.

Each time the positive half of the wave hits one of the inputs, the diode on that side pushes current toward the (+) DC output. Each time the negative half of the wave hits an input, it pulls current up from the (–) DC out. In the first half of the cycle, current gets pulled up from the bottom to the middle; in the next half of the cycle, the current gets pushed from the middle to the top. These two actions are alternating on each side of the rectifier, so that there is always current flowing up from the (–) DC out, and there is always current flowing out to the (+) DC out. This is shown in figure 15-10.

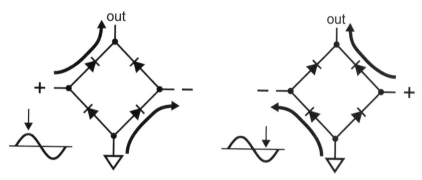

Figure 15-10.

The output of this rectifier is still not smooth, but at least it's all in one direction, as shown in figure 15-11. This is a much more satisfying and efficient solution to the rectification problem, because it harvests the energy from the half of the wave that doesn't get used in half rectification.

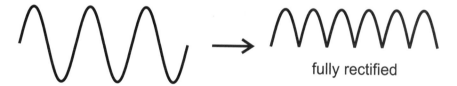

fully rectified

Figure 15-11.

On the circuit board in a piece of gear, we can use four separate diodes to make a rectifier, or we can use a premade rectifier that incorporates all four diodes into a single unit. The AC inputs and the DC outputs are marked on the case. These are shown in the photo in figure 15-12.

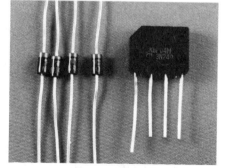

Figure 15-12.

Removing Ripple with a Capacitor

Whether we use half rectification or full rectification, we have an output that is all positive but still bumpy. We call the bumps "ripple." Technically this output is DC, because it doesn't switch back and forth between positive and negative. Any variation in the supply voltage creates noise and instability in the circuit. Ideally we would like the supply voltage to be completely stable and smooth. A capacitor is used to smooth out the rectified AC and turn it into pure DC.

A capacitor is used to remove ripple from the output of a full-wave rectifier, by placing the capacitor between the positive supply and the ground. This is shown in figure 15-13. When the voltage rises above the existing charge of the capacitor, some of the current is absorbed by the capacitor and stored as charge. When the supply voltage drops below the charge of the capacitor, the capacitor releases its charge.

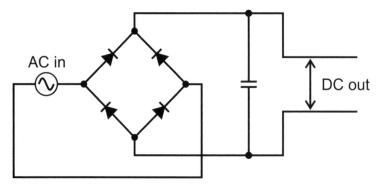

Figure 15-13.

If it's a small capacitor, this release of charge will be barely observable. If the capacitor is larger, the charge released is greater, so the voltage stays more consistent. This is shown in figure 15-14. If the capacitor is even larger, enough charge is released to keep the voltage essentially completely steady at the maximum, with very little ripple. There will always be a tiny bit of ripple, because the capacitor will always be discharging if it isn't charging. For critical applications, very large high-quality capacitors are used.

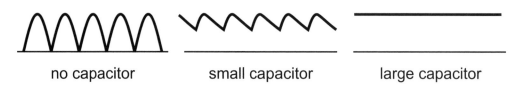

Figure 15-14.

A capacitor can also be used to smooth out the output of a half-wave rectifier. The only difference is that a substantially larger capacitor is needed, because the time between the peaks that recharge the capacitor is twice as long. This is shown in figure 15-15.

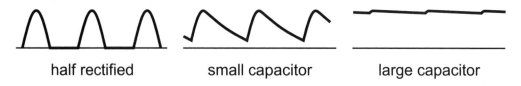

Figure 15-15.

There's an important safety point that bears repeating several times. In high-voltage supplies (as for tube amps and cathode ray tubes like older televisions and computer monitors) the capacitors used in the power supply circuit will be charged up to the supply voltage. By definition, a capacitor is a place where charge is stored, so even when the circuit is turned off and unplugged, the charge can still exist in the capacitors. The charge held on a 10,000µF capacitor charged to 300V is definitely enough to give you a very painful shock. Make absolutely sure that capacitor is discharged before working with it. Good circuit design will include bleeder resistors that let the charge dissipate when the power supply is turned off.

The Voltage Regulator

The chain of transformer, rectifier, and capacitor all serve to take high-voltage AC and bring it down to low-voltage DC. For example, we can use a transformer-rectifier-capacitor chain to take 120V AC down to 12V DC. Unfortunately it doesn't provide a consistent DC voltage output if the input AC voltage changes up or down. If the AC voltage source dropped to 100V, the output of the chain would drop to 10V. For some electronics this would be fine, but for audio we need a consistent supply. The solution is a *voltage regulator*. This component looks like a transistor, with three leads, as shown in figure 15-16.

Figure 15-16.

The voltage regulator's function is to take an input DC voltage and provide a very consistent DC output voltage. For example, to maintain a steady 12V voltage, we would use a transformer, a rectifier, and a capacitor to provide a 16V supply as the input to a 12V voltage regulator. Even if the 16V supply drops to 14V or 13V, the regulator will still maintain the level of 12V at its output.

There are many fixed and variable voltage regulators available. The most frequently seen are the fixed-voltage 78xx positive regulator series and the 79xx negative regulator series, where the numbers xx are the output voltage. For example, the 7812 has a +12V output, and the 7905 has a –5V output. Many pieces of equipment use these packaged voltage regulators to maintain consistent supply voltages in their circuit boards. Voltage regulation can also be performed by other circuitry, but the packaged regulators are easily available and easy to use.

Audio Signal Transformers

In addition to changing the voltage level of supply line voltage, transformers are also used to change the levels of audio signals. In addition to bringing the signal level up or down, audio transformers also change the impedance of the signal. A microphone uses an internal transformer to bring the signal impedance to a level where it works well with the preamplifier. In turn, older-style preamps use transformers at the input stage to adjust the signal level and impedance so that they can be amplified by the gain stage. Increasingly, preamps are being designed to eliminate the need for these transformers, but every design improvement results in a compromise being made. The best-known manufacturers of audio transformers are Jensen and Lundahl. Audio transformers are also used at the outputs of amplifiers, particularly tube amplifiers, to adjust the signal impedance to suit the comparatively low impedance levels of the loudspeakers.

Hands-On Practice

The goal of this exercise is to build a half-wave rectifier and a full-wave rectifier. We will observe their behavior by measuring and comparing their AC inputs and DC outputs.

Materials

These are the materials you will need for these exercises.

- Digital multimeter (DMM)
- Breadboard
- AC output transformer, 120V primary, 9-18V secondary
- GFCI outlet
- Four 1N4002 diodes
- 10kΩ resistor
- 0.33μF polyester capacitor
- 1μF electrolytic capacitor
- 10μF electrolytic capacitor
- 100μF electrolytic capacitor
- Two jumper leads with alligator clips
- Solid insulated wire (20, 22, or 24 gauge)

Confirming That the Transformer Is Suitable

The following exercise will only work if you have an AC transformer. Before you start taking it apart, confirm that you have the right one for the job.

Figure 15-17

- Check that the rating plate of the "wall-wart" transformer shows that its output is somewhere between 9V and 18V AC. An example with a listed output of 9V is shown in figure 15-17.

- Plug the transformer into a GFCI outlet.

- Set your DMM for measuring AC voltage on the 200V scale.

- Place one probe on each metal contact in the output plug. One probe should fit inside the inner sleeve or make contact with it, while the other probe contacts the outside sleeve. An example is shown in figure 15-18, although your output plug may be a different configuration. Because you are measuring AC, which probe is red and which is black doesn't matter, as you will get the same reading in either case.

Figure 15-18

- You should see a reading approximately equal to the voltage listed on the rating plate. Don't be concerned if the actual voltage observed is several volts higher, as this is typical. For the transformer shown in the figure, the AC voltage measured is 10.1V, even though the output was rated at 9V. It is expected that the voltage of a transformer output will drop when it is under load, so an unloaded transformer is often designed to have a slightly higher output. If you are not seeing the expected voltage, check that your meter is set for AC voltage (with the squiggly line) and not DC voltage (with the straight line).

Preparing the Transformer for the Breadboard

If the transformer is found to be suitable, you need to remove the plug.

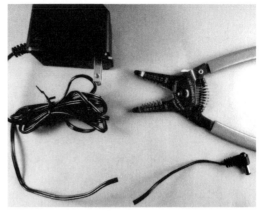

- If it is still plugged in, unplug the transformer.

- Cut the transformer cable several inches away from the plug. You can save the plug and put it back onto the transformer later if you need an AC supply for something. This is shown in figure 15-19.

- Pull the two wires of the transformer cable apart, and cut about 1" off of one of the wires. This is to reduce the chance that the two wires will contact each other when the transformer is plugged in, which usually ruins the transformer.

Figure 15-19.

- Use wire strippers or a knife to strip about 5/8" (about a finger's width) of insulation from each of the wires. The wires from the transformer are stranded, with many pieces of thin wire, so be very careful to avoid cutting through these thin wires when stripping them. This is shown in figure 15-20.

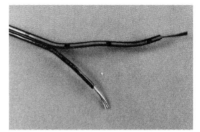

Half-Wave Rectifier Procedure

Figure 15-20.

In this exercise, you will use a single diode as a rectifier for the transformer's AC voltage output.

- If your battery clip is mounted in your breadboard, check that the battery has been removed. Leaving the battery in place during these exercises could result in the battery getting overheated, leaking, or even exploding.

- The circuit for the half-wave rectifier is shown in figure 15-21.

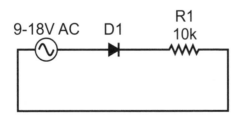

Figure 15-21.

- Use the resistor color code to identify a 10kΩ resistor, which will be used for R1.

- Insert the diode anode and cathode leads into two component strips. The completed circuit is shown in figure 15-22.

- Connect R1 to the cathode (silver-banded end) of the diode D1. Insert the other end into an unused component strip.

- Start with the transformer unplugged.

- Use a jumper lead with alligator clips to connect one wire to the anode of D1. Use a second jumper lead with alligator clips to connect the other wire to the unconnected end of R1. These are shown in the completed circuit in figure 15-22.

- Plug the transformer into the GFCI outlet.

- Switch your DMM to read DC voltage on the 200V scale.

- Read the DC voltage across the resistor, with the red probe closer to the diode. You should get a reading of less than half of the rated AC voltage of the transformer. For the 9V transformer shown, I read a DC voltage of 4.1V. This is shown in figure 15-23.

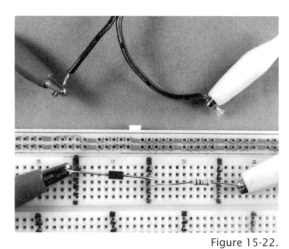

Figure 15-22.

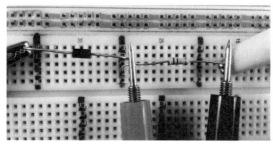

Figure 15-23.

- If we examine the AC voltage of the transformer with an oscilloscope, we see a display that looks like that shown in figure 15-24. This is a 60Hz sine wave.

- If we examine the voltage across the resistor with an oscilloscope, we see a display that looks like that shown in figure 15-25.

- Using the capacitor codes from the earlier chapter, identify the 1µF polyester capacitor. You will use this as C1 in the circuit below.

- Connect the capacitor in parallel with the resistor R1, as shown in figure 15-26. You can use a piece of wire to make these connections, as shown in figure 15-27.

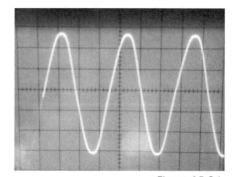

Figure 15-24.

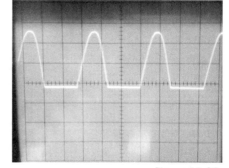

Figure 15-25.

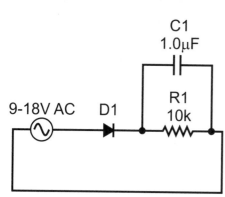

Figure 15-26.

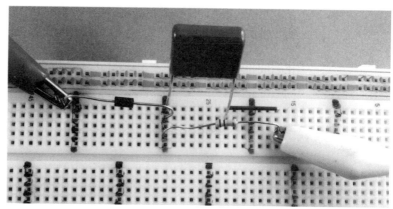

Figure 15-27.

- Read the DC voltage across the resistor, with the red probe closer to the diode. If your transformer is 9V AC, you should get a reading of about 8.5V DC. Even though the maximum voltage of the rectified wave has not changed, the DC voltage has increased. The oscilloscope shows us that by adding in the capacitor, we have filled in some of the gaps between the peaks. This increases the average DC voltage, which is what we are measuring.

- If we look at the voltage across the resistor with an oscilloscope, we see a display that looks like that shown in figure 15-28. Here we see the capacitor being charged by the rising voltage, followed by the characteristic slope of the capacitor's discharge after the peak. Notice that this discharge slope looks just like the capacitor discharge curve that you drew during the capacitor exercise.

- Identify the 10µF electrolytic capacitor. You will use this as C1 in the circuit below.

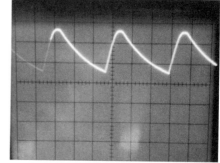

Figure 15-28.

- Connect the capacitor in parallel with the resistor R1, as shown in figure 15-29. The leads aren't long enough to reach, so use a piece of wire to complete the connection, as shown in figure 15-30. Be careful to place the (+) lead toward the diode D1. In the figure, you can see the lighter-colored band on the top of the can indicating the (−) terminal.

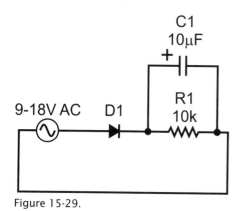

Figure 15-29.

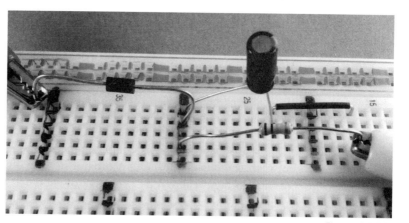

Figure 15-30.

- Read the DC voltage across the resistor, with the red probe closer to the diode. If your transformer is 9V AC, you should get a reading of about 13V DC.

- If we examine the voltage across the resistor with an oscilloscope, we see a display that looks like that shown in figure 15-31. The variation is much less than what we saw with the previous capacitor, because it is larger and so discharging much more slowly.

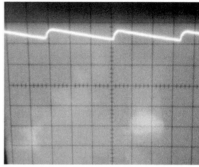

Figure 15-3

- Identify the 100μF electrolytic capacitor. You will use this as C1 in the next step.
- Connect the capacitor in parallel with the resistor R1, as shown in figure 15-32. This is just like connecting the previous electrolytic capacitor. Be careful to place the (+) lead toward the diode D1.

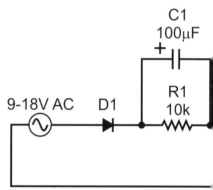

Figure 15-3

- Read the DC voltage across the resistor, with the red probe closer to the diode. If your transformer is 9V AC, you should get a reading of about 13.5V DC.
- If we examine the voltage across the resistor with an oscilloscope, we see a display that looks like that shown in figure 15-33. Even though this capacitor is even larger, there is still a little variation.
- Unplug the transformer from the GFCI.
- Clear the breadboard for the next experiment.

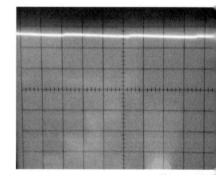

Figure 15-3

What we saw here is that we start with half-rectified voltage, which the progressively larger capacitors act to smooth out. Even with the largest capacitor used, the DC voltage was not entirely flat. An even larger capacitor would work to smooth out the output and keep it consistent. We also saw that as larger capacitors are used, the DC voltage increases. This is because the capacitor charging and discharging fills up the gaps of reduced voltage in between the peaks. By using a capacitor, the voltage is being increased to approach the maximum voltage reached by the wave of each cycle.

Full-Wave Rectifier Procedure

For this exercise, you will build a full-wave rectifier from four diodes. You will then use capacitors to smooth out the output of the rectifier.

• The circuit for the full-wave rectifier is shown in figure 15-34.

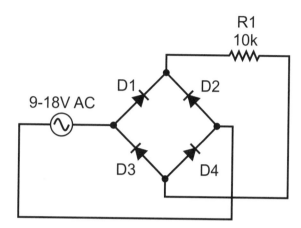

Figure 15-34.

• Assemble the four diodes as shown in the circuit, all pointing toward the top of the breadboard. The completed circuit is shown in figure 15-35.

• Add wires from the top and bottom of the diamond to the side. Add the 10kΩ resistor to connect these two points.

• Use jumper leads with alligator clips to connect the transformer wires to the sides of the diamond.

• Set your DMM to read DC voltage on the 200V scale.

• Measure the voltage across the resistor. If your transformer is 9V AC, you should get a reading of about 8V DC.

• If we examine the voltage across the outputs with an oscilloscope, we see a display that looks like that shown in figure 15-36. This is fully rectified AC voltage.

Figure 15-35.

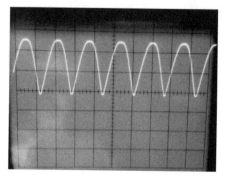

Figure 15-36.

- Identify the 0.33µF polyester capacitor. You will use this as C1 in the circuit shown in figure 15-37.

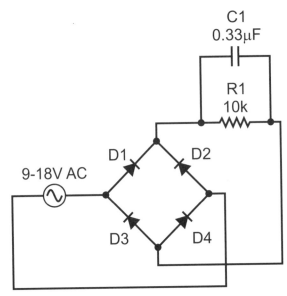

Figure 15-37.

- Add two pieces of wire from the ends of the resistors for making a parallel connection. Connect the capacitor C1 in parallel with the resistor, as shown in figure 15-38.
- Measure the voltage across the resistor. If your transformer is 9V AC, you should get a reading of about **9V DC**.
- If we examine the voltage across the outputs with an oscilloscope, we see a display that looks like that shown in figure 15-39. Notice that there is just a little bit of capacitor discharge curve visible at the bottom.

Figure 15-38.

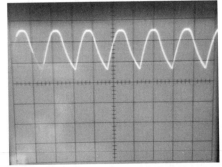

Figure 15-39.

- Identify the 1µF polypropylene capacitor. You will use this as C1 in the circuit.
- Connect this capacitor C1 in place of the 0.33µF capacitor in the circuit.
- Measure the voltage across the resistor. If your transformer is 9V AC, you should get a reading of about 11V DC.
- If we examine the voltage across the outputs with an oscilloscope, we see a display that looks like that shown in figure 15-40.
- Identify the 10µF electrolytic capacitor. You will use this as C1 in the circuit.

- Connect this capacitor C1 in place of the 1μF capacitor in the circuit. Be careful to place the (+) lead toward the top of the rectifier. This is shown in figure 15-41.

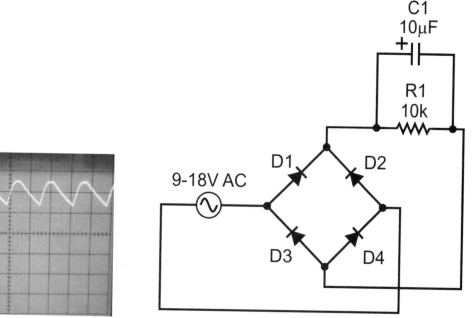

Figure 15-40.

Figure 15-41.

- Measure the voltage across the resistor. If your transformer is 9V AC, you should get a reading of about 12.5V DC.

- If we examine the voltage across the outputs with an oscilloscope, we see a display that looks like that shown in figure 15-42. We still see a little variation in the voltage, but there is much less than there was before.

- Identify the 100μF electrolytic capacitor. You will use this as C1 in the circuit.

- Connect this capacitor C1 in place of the 10μF capacitor in the circuit. Be careful to place the (+) lead toward the top of the rectifier.

- Measure the voltage across the resistor. If your transformer is 9V AC, you should get a reading of about 13V DC.

- If we examine the voltage across the outputs with an oscilloscope, we see a display that looks like that shown in figure 15-43. The voltage supply with this capacitor is essentially flat, with no visible ripple. Note that when this same capacitor was used with the half-rectified supply, it was not large enough to remove the variations in voltage. This is one reason to prefer a fully rectified

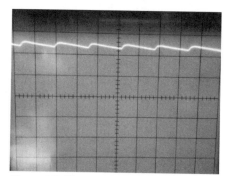

Figure 15-42.

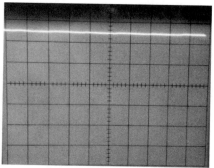

Figure 15-43.

supply to a half-rectified supply. Both can have the ripple removed with a capacitor, but half rectification requires a much larger capacitor.

- Unplug the transformer from the GFCI.
- Clear the breadboard.

This exercise has shown the roles of the transformer, the half-wave rectifier, the full-wave rectifier, and the capacitor in bringing large AC voltage down to usable levels of DC voltage. It has also introduced the ability of the oscilloscope to visualize voltages that are changing over time.

16

Safety and Test Equipment

Goal: When you have completed this chapter, you will be familiar with the equipment needed for testing audio gear, and with the precautions needed to work safely while using this test equipment.

Objectives

- Identify strategies that are used to disassemble gear for testing to allow consistent reassembly.
- Identify the dangers of working with electricity.
- Recognize and put into practice the measures needed to work safely.
- Identify the situation in which to use a Variac during testing.
- Identify the testing capabilities of a digital multimeter (DMM).
- Identify the purpose of a function generator.
- Identify the testing capabilities of an oscilloscope.
- Recognize the configuration of function generator and oscilloscope used to test the operation of a piece of audio gear.
- Identify the purpose and advantages of a cable tester.

The Essentials of Working Safely with Electricity

Several hundred people die of electrocution every year in the US. Don't become part of that statistic! Always remember the First Rule of Electricity, which is that it's trying to kill you all the time. The amount of current that can kill you is fairly small. 300mA across the heart will do the job. This is just three-tenths of an amp. Most circuits are on circuit breakers that are rated at 15 or 20 amps, and they don't trip fast enough to save you in any case. Those circuit breakers aren't there to save you from the supply line voltage. They are there to keep the building from burning down as the result of an electrical fire. The only way you can stay alive while working with electricity is to be extremely cautious when you're working with it.

The following guidelines are by no means the only instructions you need to work safely, but they should cover most situations. The most important point is to be patient and thoughtful. The vast majority of bad experiences I have had could have been avoided by taking on the project a little more slowly.

- Don't routinely work on equipment that is powered up. Sooner or later you'll need to power it up while you've got it open so you can see what's wrong, but do this only under very controlled circumstances and with the right equipment. If you need to unscrew something, open something, pry something off, or anything else of this nature, power down and unplug the unit! Avoid the temptation to say, "Oh, it will be okay just this once. I know what I'm doing." It may be true that one time, but it's also the beginning of continued unsafe practices.

- Don't trust the power switch to turn the equipment off. Old-fashioned physical power switches do in fact turn off the voltage supply to the rest of the circuit, but the voltage supply line is hot up to the switch, so there is still danger inside the enclosure. Newer equipment is on all the time, even when it is turned off; the voltage supply is not actually connected to the power to the power switch. Instead it is routed through a relay or other device that is turned on and off by a low-voltage signal. This means that so long as the unit is plugged in, there is supply voltage on some part of the circuit board.

- To be safe with old or new equipment, turn off the switch and unplug it from the socket. When you're working, have the socket you are using in clear view. Get in the habit of dangling the plug over the edge of the work surface and looking at it to make sure it's out of the socket before you do anything else. Certainly this is a hassle if you're doing lots of testing, but it's much better than getting electrocuted. A power strip with an on-off switch is useful for testing, but you still need to unplug the unit from the strip to be safe.

- Logically, if the power is disconnected, we expect that the equipment is safe. Nothing could be further from the truth. Some pieces of equipment, such as power amps and tube amps, have large power capacitors that can hold a substantial charge for a very long time (days or weeks). The charge slowly bleeds off through the resistance of the rest of the circuitry. Tube amps use a DC power supply of about 300V DC. Even more dangerous is anything with a cathode ray tube, like a TV, video monitor, or old (non-flat-screen) computer monitor because the second anode on a CRT is charged to several thousand volts. Don't work on this equipment unless you're specifically trained to. Just waving your screwdriver too close to the wrong thing can give you a really nasty shock. Fortunately, real CRTs are becoming increasingly rare, so tube amps and power amps are the biggest dangers.

- The safe way to test a unit that uses supply line AC voltage from the electrical outlet is with an *autotransformer*, which is also called by a common brand name of Variac. This transformer allows you to regulate the 120VAC supply line voltage from 0VAC up to 140VAC, as shown in figure 16-1. If you have a piece of equipment that smokes or shoots sparks when it's plugged in, you can plug it into the Variac, turn it on at 10V, and troubleshoot relatively safely. This voltage isn't enough to harm you, and often allows you to find the problem without any fireworks. You can pick up a used Variac on eBay for $50 or so.

Figure 16-1.

- If you don't have a Variac, you can do individual tests of the points you want to look at. While the unit is still unplugged and turned off, set up your DMM to read the voltage you want, and use patch cables with alligator clips to attach it to the points you want to test. Get your hands out of the unit, turn on the unit, step back, plug it into your power strip, and turn on the power strip. Look at the DMM, and write down your reading. If you need a reading in a different place, go through the whole process again: turn off the unit, turn off the power strip, unplug, move the alligator clips for the probes, plug it back in, turn on the power strip, turn on the unit. This is a slow process, but you get to stay alive to retain the knowledge you are gaining about what you're fixing and to enjoy other pleasures of living also. Hurrying the troubleshooting process when working with high voltage can be deadly. Once you've been doing it this way for several years, you'll have a better sense of where it's safe to put your hands and your DMM probes, and you can start taking a few short cuts. Until then, it's better to treat every piece of audio gear with the same respect you would give to a poisonous snake.

- With or without a Variac, when working with a scope probe or DMM inside a powered-up unit, be extremely conscious of where your hands and probes are at all times. There's a reason those probes are insulated! If you're concentrating on your right hand, keep your left hand out of harm's way. It's best to keep one hand in your pocket or on your lap while working inside a piece of live equipment. That means not touching anything metal or anything grounded. Of course you'll need both hands for both DMM probes, but in a lot of situations you can get away with just using a clip to attach the probe to signal ground.

- If possible, work with a spotter when you're working inside hot equipment. Make sure that your spotter knows the address where you both are, has access to a telephone, and knows how to dial 911. It's even better if your spotter knows CPR. It's not so good if your spotter is a distraction from the task at hand. The best spotter is one who sits and does something else quietly in the same room while keeping an eye on you as you work.

- When a piece of equipment is plugged in, only one person at a time should be working on it. It's hard enough keeping track of your own left and right hands without having to think about someone else's hands also.

- If you need to solder something inside, unplug it while you're working. Never, ever solder on a live board! It can be detrimental to you, your soldering iron, and the circuit you are working on.

- Equipment with a 3-prong plug is grounded, which is safer than ungrounded, but it's not a complete safeguard. It's still entirely possible to get a deadly shock from this equipment. The earth ground is only effective at protecting you in certain situations.

- If you're setting up a bench where you're expecting to test equipment, equip it with a GFCI circuit. As discussed in an earlier chapter, the GFCI will turn off the supply voltage instantly if it detects that you have completed the circuit to ground. Just because a piece of gear is hooked up to a GFCI doesn't mean you are completely protected. The capacitors mentioned earlier remain charged (and able to discharge lethally) even when the power is completely turned off and disconnected, so a GFCI that turns off the power isn't going to help if you touch them.

The Digital Multimeter (DMM)

The DMM is the first-line tool in looking at a problem in the studio or on a piece of audio gear, and we've already seen how we can use it in several ways. This meter is several different meters combined into one unit, namely a voltmeter, an ammeter, an ohmmeter, and typically a continuity tester or diode tester. The ohmmeter and continuity functions can be used to test continuity, which is whether two points are connected that should be. These same tests can assess whether there is a short, which is when two points are connected when they shouldn't be, such as the shield and the conductor of a cable. The voltmeter can be used to measure AC voltage and DC voltage. The AC voltmeter function can be used to determine if an electrical socket is live. In some cases, the AC voltage function can also be used to check the signal output. Once the cover is off of a piece of equipment, the AC and DC voltmeter functions can be used to confirm that appropriate levels of voltage are present in the expected places on the circuit board. More complex diagnostics can also be performed with a DMM, but these are the typical or basic functions.

A serviceable DMM costs $30–$40. Wavetek is a reasonable brand. Fluke is an excellent brand (and is consequently much more expensive). There are professional bench models that run in the thousands of dollars, but I don't have one on my own bench. There are even little meters the size of a pocket calculator that work fine for most applications. They don't usually measure current, but most field testing uses the voltmeter, ohmmeter, and continuity tests, so current measurement isn't needed.

The Oscilloscope

The oscilloscope is the workhorse for troubleshooting equipment. It is also used for aligning the heads of analog recording gear. Its function is to graphically display any AC voltage that is changing as a function of time. Briefly, the oscilloscope probe is touched to the conductor that is carrying the signal. The signal is amplified by the oscilloscope and displayed as a point of light moving up and down on a small screen. At the same time, the timebase controls can be set to move the point of light from left to right across the screen. When it goes off the right side, it starts over on the left side. When these two movements are combined, a wave pattern is created on the screen, as shown in figure 16-2. A typical analog oscilloscope is shown in figure 16-3.

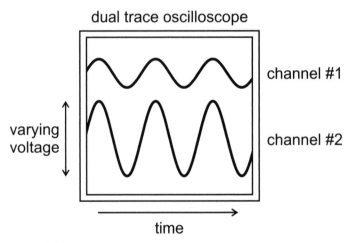

Figure 16-2.

The sensitivity of the probe can be varied over a wide range, usually a thousand-fold or more, so that everything from 120VAC from the wall socket down to the tiny signal of a few millivolts from a microphone capsule can be displayed. By altering the speed of the scan, the entire audio spectrum can be covered. On more advanced instruments, much faster signal oscillations (such as RF, radio frequency) can be displayed.

Paired with a function generator, the oscilloscope can be used to show how a signal is making its way through a piece of equipment, what is happening to it along the way, and how the equipment's controls are affecting it. Signal clipping is graphically displayed, as are changes in tone and phase. With the oscilloscope, it is possible to trace a signal through a circuit until you find where it's being blocked, degraded, or lost. Once the problem component has been identified, it can be removed and replaced.

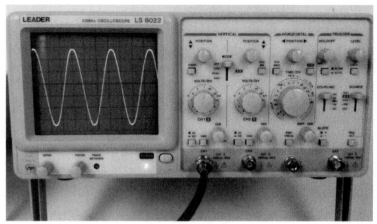

Figure 16-3.

A used analog oscilloscope of reasonable quality can be purchased for $50–$150 on eBay. A used digital oscilloscope (also called a storage scope) can be found for $400–$800. Packages for using a computer as an oscilloscope display and controls are also available new for about the same price. While it's attractive to use a computer for everything, I still prefer the analog scope, and not just because of the lower price. There is something pleasing to me about being able to control the display with physical knobs and switches.

Function Generator

The function generator can be used as the source of a test tone that is run through a piece of equipment to confirm its output or observe what kind of effect the equipment is having on it. Typically an oscilloscope is used to display both the original signal and the output from the equipment so they can be compared.

A typical function generator is capable of producing precise signals of several waveforms. These are typically sine and square waves, and more rarely triangle waves. In specific situations, pulse waves are useful. These waveforms are shown in figure 16-4.

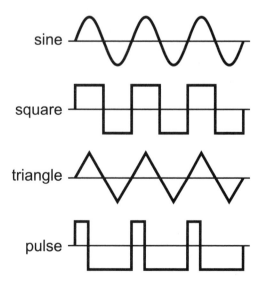

Figure 16-4.

A typical function generator is shown in figure 16-5.

Figure 16-5.

Most function generators maintain the same frequency of wave once they have been set. Some function generators can be set to sweep continuously through the spectrum of frequencies between lower and upper limits. These sweep generators are used to assess the functioning of a circuit or a piece of equipment in a particular frequency range. They are much more convenient for this purpose than changing the output frequency by hand.

Each of these waves has a diagnostic function when used with the oscilloscope. For example, the square wave is useful for assessing the effects of tone controls and similar frequency response issues, while the sine wave is better for observing amplification and clipping.

A used function generator of good quality can be purchased for $50–$75 on eBay. You can also build one from a kit, but it's far easier to buy one.

The Cable Tester

The cable tester is a piece of equipment that simplifies the testing of cables and connectors for continuity and shorts. This small handheld device is invaluable for quickly and easily determining if a cable or its connectors are faulty. You can do the same tests with a DMM, but it takes longer and you have to think more. LEDs on the cable tester indicate whether the signal path between two connectors is good. A typical cable tester will have male and female XLR jacks (for testing mic cables and other low-impedance cables), two 1/4" phone jacks (for checking TS and TRS cables), and two RCA jacks. Some also allow you to test TT cables, MIDI cables, and optical cables. Some cable testers include a test tone generator and can test the presence and level of phantom power. Cable testers are relatively inexpensive, especially since the advent of low-cost electronics equipment manufactured overseas.

Introduction to Troubleshooting and ESD Precautions

Goal: When you have completed this chapter, you will be familiar with the basic methods of solving the problems of a piece of audio gear that is not working properly.

Objectives

- Identify the dangers ESD poses to audio gear.
- Understand how to avoid ESD damage.
- Recognize the symbol for a fuse.
- Use the DMM's continuity function to test a fuse.
- Use the DMM to test a circuit board for supply voltage.
- Distinguish between problems of voltage supply and problems of function.

Electrostatic Discharge

A charge is developed on an object by bringing it into contact with another object and then separating them. This is called a *triboelectric charge*. A classic example is shuffling your feet on shag carpet. When you move your feet, you are making and breaking thousands of contacts with the fibers of the carpeting. Some of the electrons on the fibers get stripped off and hang onto you instead of staying with their original atoms where they belong. (Note that there are no recording studios with thick shag carpet!)

Now you've got a charge on you. In essence, you're acting like a capacitor. While the number of electrons is quite small, their potential can be very high, such as several thousand volts. The charge wants to go somewhere, to get back to the ground it was so roughly separated from. If the air is moist, the electrons will jump off your body onto the tiny water particles that are floating around in the air. If the charge doesn't dissipate on its own, it just waits around. The definition of the word *static* is to not be going anywhere, to be immobile, or to be unchanging. At this point the charge on your body is static, because it's not going anywhere else yet.

If you touch something that allows the electrons to get back to ground, an *electrostatic discharge* occurs. Electrostatic discharge often gets abbreviated to ESD. It's a short pulse of very high voltage with very low current. If you touch a doorknob, it's no big deal because the doorknob doesn't suffer from a burst of high voltage. If the discharge goes through a piece of sensitive electronics on its way to the ground, it can seriously harm the semiconductor components. It could kill the equipment altogether or degrade its performance. Some types of components are very sensitive to ESD. Theoretically the chassis should take the brunt of ESD, but if you're working with the cover off or in a very dry environment, ESD can get in and damage the components on the printed circuit boards.

There are all sorts of complex and expensive ESD control products and procedures that are used in electronics assembly plants. Most of them don't apply to the kind of work we do with fixing electronics. There are a few precautions that we need to take to keep our gear safe while we are working on it.

- Use a wristband plugged into or attached to a grounding point that connects you to the earth ground.

- Attach the equipment's chassis ground to that point also.

- Work on an anti-static mat if possible.

- If no earth ground point is available, use a wristband to attach yourself to the chassis ground. This is less than optimal because although the electrostatic charge doesn't build up on you, any charge you create will find its way back to ground through the equipment.

- Keep sensitive components in ESD-safe bags or other containers until use, or keep extremely sensitive components wrapped in foil or with their pins stuck into conductive foam as well as in ESD-safe containers; this prevents one pin from being brought to a higher potential than another.

- Don't touch the signal-carrying conductors of a connector when the other end of the cable is plugged into a piece of gear.

Troubleshooting a Piece of Equipment

As discussed in the previous chapter, the first rule of working with electronics is to keep out of the path between a high voltage source and the ground.

By its very nature, each piece of equipment in the studio has a limited time that it will continue to function. Most gear has hundreds of components, and sooner or later one of them is going to fail. If you know what you are doing, you can figure out which component has failed, get a replacement, and fix it. It's also possible that you'll cause more problems, which you may or may not be able to fix. This is just part of the learning curve. As annoying as it may be, remember the second rule of working on electronic gear: know when to stop! Sometimes the point to stop was five minutes earlier than you did. When this happens, figure out if you can undo the damage. It's frustrating, but it's also a learning experience. If you try everything you know how to do and nothing helps, it's time to ask for help or to take it to someone more knowledgeable who can fix it for you. I've done a fair amount of both of these.

Disassembly Tips

If you get into troubleshooting equipment at all, you'll have to take things apart. There's a big problem with disassembly. Often, when it comes time to put things back together, you find that you don't quite remember where everything goes. As you are taking things apart, it's really easy to think that you will remember where everything goes back. Unless you have opened the unit several times before and you really know your way around inside, when you start removing fasteners you are likely to lose track. You have no idea how many screws you need to take out after the first one, whether you will be removing just a few or a lot. There may also be cables and internal connectors to disconnect. These absolutely must be put back in the right places when you are done. If you have a photographic memory, you'll have no problem. The rest of us find that taking frequent pictures with a digital camera or camera phone as we're taking things apart really help when we go to reassemble. Take extra pictures, because it is surprising how often the first photo doesn't quite show what you really need to see, or it's a little blurry or a little off center. If you don't end up needing them, you can just throw them away.

Another trick is to label each screw as you take it out and to label the hole it came from. For labeling you can use the convenient books of numbered strips for numbering electrical wiring that are sold in the electrical section of hardware stores. You can use a 28-compartment pill container to put the screws in as you remove them, or line up the screws and other fasteners and number them and stick them down to a piece of cardboard. The advantage of the pill containers is that you are less likely to drop them all on the floor and lose them, as always seems to happen when working on equipment. Also number and label the connectors you remove and unplug. This is tedious to do, especially when you are impatient to get inside and figure out what the trouble is, but it's worth the effort. In the end it's much easier than trying to put everything back just from memory, especially after your project has been sitting on the bench for a few weeks or more while you are waiting for a part to arrive. There can be disastrous consequences to putting the wrong screws back in the wrong places or putting things back in the wrong order.

The Basic Steps of Troubleshooting

There are two basic classes of problems that can befall electronic equipment, namely power supply problems and function problems. It doesn't do any good to address problems of function if there are problems with power supply.

Power Supply Problems

Every piece of electronic gear needs a voltage supply of some kind. If the supply isn't connected to the rest of the circuit and isn't at the right level, the gear won't work. I once worked on a console that was exhibiting all sorts of strange problems: preamps that didn't work, channels that got stuck in solo mode, low output on some channels but not others, and EQ that wouldn't switch off. When I tested it at home, it seemed fine. The next day, all the problems were back. I traced the problem to an intermittent negative voltage supply. When I got that fixed, the console was fine.

Check the Outlet, the Battery, or the Power Supply

If the outlet is dead, you can poke around inside the gear until the cows come home and never find the problem. Use your DMM to check that there's 120VAC available at the wall socket as there's supposed to be. A reading of 105V is barely acceptable. Usually you'll get a reading of 110V or 115V. If you see a zero, check that your meter is set to measure AC voltage, as a meter on DC voltage will read zero on a working AC outlet.

If there's a battery supply, check that the battery is good. The battery voltage should be no more than 10% below the listed voltage. For a 9V battery, it should be no more than 0.9V down, at 8.1V or more. Even then, you know it's time to start looking for a replacement battery.

If there is an external power supply, use your DMM to check the voltage, as you did in the hands-on exercise with transformers. Be sure to read what you expect the output voltage to be, AC or DC. Most "wall-warts" provide low-voltage DC, but some provide AC, and you need to be on the appropriate setting. Be careful with your DMM probes to keep from inadvertently shorting out the supply by touching both contacts at once, as this can destroy a previously functioning power supply. Also check that you have the correct power supply because equipment that runs on DC won't work on AC, and vice versa.

Check the Fuses

Almost every piece of gear has at least one fuse. The fuse is a component with a wire that is designed to melt (or blow up) if more than a certain amount of current is passed through it, thus turning off the flow of current. It's a safety feature to prevent too much current passing through a circuit. Without the protection of the fuse, the unit is vulnerable to damage to the circuit, melting into a pile of slag, catching on fire, or electrocuting the user.

The schematic symbol for a fuse is shown in figure 17-1, along with a photo of some fuses. There are many models of fuses on the market. The ones we see most often are the little glass or cardboard cylinders with metal caps at each end. The glass-bodied fuses are easy to diagnose. A working fuse has a wire or metal strip visible inside, while a blown fuse will just have little bits of melted metal rattling around. The cardboard-bodied fuse is a little more difficult because we can't see inside. Fortunately it's easy to test a fuse with the DMM on the continuity setting. In the example of the fuse symbol, the designation is shown as F1, and the fuse's current limit is shown as 250mA.

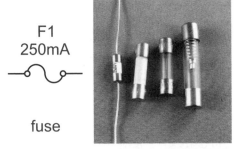

Figure 17-1.

The continuity test setting on the DMM is shown in figure 17-2. Most meters when set to this position give an audible beep or tone when there is a connection between the two probes. This meter setting is indicated by a symbol for sound, as shown in figure 17-2. It's good practice to always touch the probes together when starting to use this setting, to check that it is working properly.

Once you know that the meter is working properly, you can use it to test the fuse. This is shown in figure 17-3. If the fuse is good, the meter will give a beep. If it is blown, with no connection, there will be no beep.

Sometimes the fuses are in a convenient fuse holder in the back of the chassis, and you won't even have to open up the case. Sometimes they're inside. Sometimes they are even soldered directly to the circuit board. If you replace the fuse and it immediately blows again, you've probably got a shorted component somewhere on the circuit board, and it's time to start the diagnostic process to find it.

Figure 17-2.

If a unit is constantly blowing its fuse, it's tempting to install a fuse that has a higher limit. Don't even think about it! The fuse is there to act as an emergency safety cutoff, and the specified fuse is for the maximum current that the designers expected the unit to use, ever. Using the wrong fuse can be very dangerous, resulting in the circuit board catching fire or the user being electrocuted.

Look on the Circuit Board

The DMM is all you need to diagnose whether there is a voltage supply on the circuit board. However, you need to open up the piece of gear to get to the circuit board, so it's often easier to skip

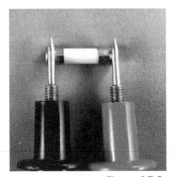

Figure 17-3.

down to checking inputs and outputs and come back to this step if there is no signal coming through. With the circuit board exposed, check that you have DC voltage in the appropriate place on the board when the switch is turned on. If you do, you can continue on to figuring out what's wrong with the inputs and outputs. If not, you need to work through the line of electrical supply to find the problem. One easy place to check the supply voltage is on the pins of any integrated circuits (ICs) on the board. Integrated circuits are combined sets of semiconductors that perform particular functions. Their most typical forms are dual inline packages, called DIPs for short. Some examples are shown in figure 17-4.

ICs are good candidates for checking the voltage supply on the circuit board, because they each require a supply voltage. Check each pin of the IC with the red probe of the DMM set as a voltmeter and the black probe on the signal ground. This is shown in figure 17-5. Be careful to only touch one pin at a time, with the other probe on the signal ground.

Remember that signal ground is easy to find in a piece of audio gear. Signal ground is pin 1 of any XLR connector, or the sleeve of any TS or TRS connector. If it's a completely transistor-based design with no ICs and if you don't have the schematic, it can be a little more challenging to figure out where you expect the voltage on the board to be. Generally one leg of each transistor should be electrically hot. Checking voltage supply on tubes is easy but dangerous, because the underside of the socket offers easy access, and the supply is typically 300V DC. Be very, very careful checking the power supply in tube equipment!

Figure 17-4.

Function Problems

If it's not a problem with the voltage supply, it's most likely to be something wrong with the electronics that do the work of amplifying, compressing, limiting, or whatever the unit does.

Define the Problem

If you've been working with the gear, and you are familiar with the normal functioning, then this will be pretty obvious. If someone else brings you a piece of gear and says, "It's not working, and Jane said you could probably fix it," find out what the actual problem is before you start taking off the cover plates.

Figure 17-5.

Check for Smoke

We've all smelled that horrible crispy-fried electronics smell when a piece of gear dies. If it starts to smoke, don't try to locate where the smoke is coming from—unplug the thing immediately! Now you know where at least one problem with the gear is. The usual culprit in this situation is a transistor or voltage regulator, so these are the first components to check. Some people use the strategy of letting it overheat so the bad part can be spotted, but this is very risky, because it often results in damage to other components on the board or to the board itself. Don't do it. Instead, use a Variac to reduce the voltage. In this situation, a low voltage such as 15VAC can be supplied so that you can safely feel around on the board (or use a temperature probe) to find out what is getting hot without it actually burning you, bursting into flame, or melting into a puddle of slag.

Compare the Input and Output

This is the big one. Most gear in the studio operates on an input-output basis. Consoles, preamps, recorders, and signal-processing gear are all expected to be altering a signal as it passes through the unit. We can start to assess whether the unit is doing anything by hooking up a signal source (a function generator) at one of the inputs and using an oscilloscope to see what we get from the output. A diagram of this setup is shown below in figure 17-6. If you see something at the output, you can ask and answer questions about frequency response and distortion. If you don't see any output, then it's possible to open it up and try to trace the signal through the circuit to find where it's being blocked.

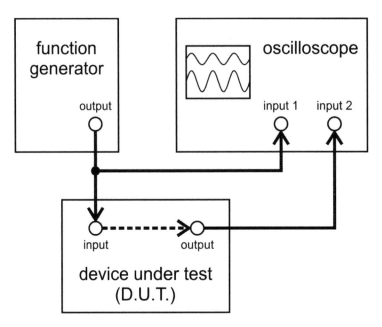

Figure 17-6.

This kind of analysis may take some time, but it can be very instructive about amplification levels, frequency response, and distortion.

Read the Schematic

Some equipment manuals include the schematic for that piece of gear. At first they just look like spaghetti, but with some practice you can identify where the inputs and outputs go and what happens to them along the way. There are a limited number of amplifying components, namely transistors, op amps, and tubes. The true work of most audio gear is done by these components. These are the parts most likely to be causing the problems if the audio signal path is dead, so they are the candidates to be checked first. The signal has to go through one of these to be amplified. Not all of the transistors or op amps in a piece of gear are involved in amplification, but many of them probably are. You don't have to be an electrical engineer to find your way through the basics of the signal flow. Try it and see. If you don't have the schematic yet, look around for it on the Internet. There are substantial libraries of equipment schematics and service manuals available online.

Check the Transistors

There are various tests that can be done to check transistors, but the easiest is just to use the continuity test or the diode test on your DMM. Make sure to do these tests with the gear unplugged. First check for DC voltage across the three contacts of the transistor, in three tests: left to center, center to right, left to right. All of these tests should show zero voltage. If they don't, wait until the capacitors discharge before continuing! Once it is safe, test the three contacts of the transistor with the continuity test of your DMM: left to center, center to right, left to right. If any of these three tests shows continuity, the transistor is bad and must be replaced. That may not be the only problem, but it's a good place to start.

Test the Op Amps

The term *op amp* is short for operational amplifier. Op amps are packaged in IC chips. Replacing an op amp may be either really easy or really hard. It's easy if the gear is older and the op amp chip is mounted in a socket. In this situation, you just need to pull out the old chip and put in a new one. With newer gear with SMT (surface mount) components, it's a little more difficult, but still achievable. Part of the trick here is how to know whether the op amp is actually part of the signal path and whether you expect signal to be going through it. Op amps are very common in audio circuitry because they are so versatile. They get used as signal-summing stages, inverting amplifiers, non-inverting amplifiers, buffers, active filters, voltage regulators, meter and indicator drivers, and comparators. Because of this, it's not always that easy to just point to one on the circuit board and say, "That's an op amp, and I should see signal output from it on pin 1. I'm not seeing any signal, so I should replace it." We need to see the op amp in its circuit context to make this set of conclusions, and that means being able to analyze the circuit and evaluate where the signal should be going through it. This process is easily the subject of a whole book. I'll just say generally that the method of determining what an op amp or other IC is doing is to study the circuit diagram and figure out where the signal is going, so that you can develop a plan of the points of where to test.

Change the Tubes

Tubes are harder to test these days, so the easiest thing to do is to switch out the suspected tube with a new one. (In the good old days, back when I was growing up and dinosaurs were still walking the earth, there were tube testers in the hardware store and even in the grocery store.) It's all too easy to lose track of which tube is the new one and which is the old one, so mark the new tubes with a Sharpie before you start. Unfortunately, new tubes are usually pretty expensive. Another alternative is to search out a local tube-gear person who has a tester.

Think Outside of the Box

Sometimes a problem that appears to be electronic in nature turns out to have a mechanical cause instead. Because electronics are complex (and because we want to try out the cool stuff we know), we tend to jump to the conclusion that something wrong with the unit is caused by component failure. Keep an open mind about this conclusion, because sometimes it is wrong. About the time you're scratching your head about what the heck this electronic problem might be, ask yourself if there's something else that might account for the situation. I once spent most of each weekend for several months trying to diagnose the electronics problem of a bass head that shocked me every time I touched the strings. The electronics were fine. The problem was in the power switch, which was just a little too large. It was almost touching the inside of the chassis, and every time I turned on the amp, the chassis became electrically charged with supply line voltage. This should have been a relatively easy problem to spot, but I was focused on fixing the electronics, so it took me a while to step back and look at the problem from a larger perspective.

18

The TS Audio Cable

Goal: When you have completed this chapter, you will have constructed an unbalanced audio cable with Neutrik TS (tip-sleeve) connectors.

Objectives

· Strip and tin the shielded cable properly for soldering.
· Tin the connectors properly for soldering.
· Solder the TS connectors to the cable with clean, shiny solder joints.
· Assemble the cable's sleeve, boot, and strain relief appropriately.
· Test the cable for continuity and shorts.

Overview of the Soldering Projects

These projects are designed to familiarize you with the construction and function of three common audio industry cables and a piezo transducer. All of the materials needed for these four projects are listed together in the front of the book, along with specifications and suggested sources.

The way I build cables is not the only way. The bottom line is whether the cable works or not. I follow certain work habits to build cables that are durable, long lasting, and trouble free. Some of the steps described here may seem unnecessarily picky. I've developed these habits over many years. You should feel free to experiment with different methods. I won't be looking over your shoulder and telling you that you are wrong. With that said, I recommend learning the methods described here and then finding what shortcuts you think you can take. These methods may feel slow at first, but with experience you can make a cable in about 15 or 20 minutes. It just takes practice.

The tip-sleeve audio cable or TS cable is often called a guitar cable. It is one of the fixtures of the audio industry, both for electric guitars and for keyboards. Its origin is in the telephone industry, back when there were hand-operated switchboards. For this reason, the TS connectors are also called phone plugs. This is an unbalanced cable, with one conductor for the signal and a shield conductor for the ground. Because of this single conductor for the signal, we also call it a monaural or mono cable. This distinguishes it from the balanced or stereo cable, which has two separate signal conductors.

Materials

These are the materials you will need to build the TS cable with Neutrik connectors. Connector and cable specifications and sources can be found in the introduction.

– Two TS connectors, model Neutrik NP2C
– Unbalanced heavy-duty cable
– Safety glasses
– Soldering station or soldering iron
– Solder
– Helping Hands jig
– Panavise or similar vise
– Reamer/scraper tool
– Wire strippers
– Needle nose pliers
– Craft knife (X-acto or similar)
– Digital multimeter (DMM)

The Neutrik TS Connector

For a long time, the connectors typically used for high-quality guitar cords, and other audio cables were made by Switchcraft. While these are good-quality connectors, they do not have any built-in strain relief. Some connectors use a coil-style strain relief, which can help prevent damage to the cable from being bent too tightly, but these do not actually relieve any stress on the cable's joint with the connector. A well-soldered joint is usually up to the task, but in commercial situations (such as being used and abused in the studio or on the road), a connector with built-in strain relief will last longer and be more reliable.

The current industry standards for high-quality audio connectors are those made by Neutrik. As well as being manufactured without any rivets, they have strain relief built in, so that when completely assembled, there is no strain placed on the solder joint inside. The Neutrik NP2C phone plug is shown in figure 18-1. These connectors are somewhat more challenging to solder than their Switchcraft counterparts.

Figure 18-1.

The Connector Parts

The NP2C comes packaged as four parts in a plastic bag. The parts are the body, the sleeve, the strain relief, and the boot. If you have not worked with a particular connector before, it's a good idea to dry assemble it and take it apart again before you start soldering. This short procedure will familiarize you with the parts and their relationship to one another, and it may save you from arriving at the moment of final assembly only to find that you should have slid a part over the cable before you started soldering. The four parts are shown on the left in figure 18-2.

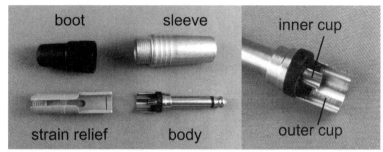

Figure 18-2.

The cable shield gets soldered into the inside of the outer rim of the body, the outer cup. The inner core wires get soldered into the center cup. These parts are shown on the right in figure 18-2.

The four parts of the connector work together to keep the strain of the cable off of the solder joint, so that the outer insulation takes the strain if the cable gets pulled on. The fingers of the strain relief clamp down on the outer insulation to hold it. The sleeve screws onto the boot to press the body of the connector into the strain relief. This compresses the fingers of the strain relief onto the cable and holds it tightly.

The strain relief clamp is designed to work with a wide range of cable diameters. For thin cables, the fingers of the clamp are pressed farther together, so the clamp needs to be a little longer. For thick cables, the clamp needs to be a little shorter, or the sleeve won't be able to tighten far enough onto the threads of the boot. The strain relief is designed for this situation by having a C-shaped spacer at the end of the extension. On thick cable, this extension is removed before putting the strain relief on the unstripped cable. On thin cable, the extension is left in place.

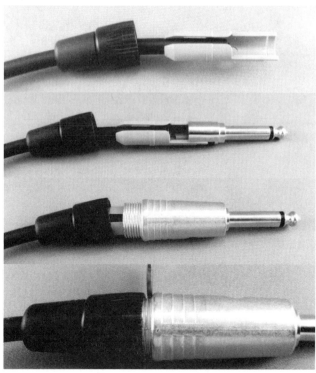

To check whether to remove the spacer or not, you can do a dry assembly of the cable and connector, as shown in figure 18-3. Put the boot on the cable, and slide the cable end halfway into the strain relief. Assemble the connector with the body and the sleeve. When the sleeve is tightened, there should be only a small area of threading visible. With the sleeve fully tightened, try pulling on the cable. You should not be able to be pull it out. If the gap of exposed threading is wider than double the thickness of a penny, as shown in the last panel, the spacer should be removed.

Figure 18-3.

The strain relief with the spacer removed is shown in figure 18-4.

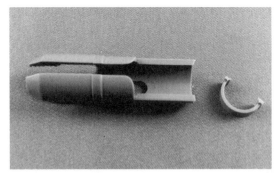

Figure 18-4.

The Cable Parts

The cable you will use has a single conductor with a shield. This is also called mono cable. The single bundle of stranded wires at the core is for the signal, and the braided wrapping of wires around the outside is the shield. The layers of the cable are shown in figure 18-5.

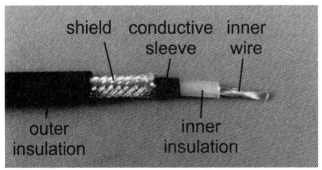

Figure 18-5.

On the outside is the outer insulation. This rubber or plastic is formulated for good resistance to abrasion. We want thick outer insulation in the studio and for live sound. Good insulation allows the cable to keep functioning after being stepped on, having the edges of heavy stage cases rested on it, being rolled over with a camera dolly, and all the other abuse that we often subject it to.

Next is the braided shield. This is a network of thin wires braided into a crisscross pattern. The function of the shield is to absorb background hum and hiss and other undesirable electromagnetic radiation. The braided pattern is used so that there are fewer holes in the shield when the cable is flexed or twisted. For best results, never use cable that doesn't have either a full braided shield (such as the recommended Redco TGS-HD and Canare GS-6) or a high-quality unbraided shield (such as Mogami W2524) combined with a conductive inner sleeve. Unbraided shield without the conductive sleeve is less expensive, and it's easier to work with, but it doesn't provide the hum rejection of a professional cable.

Next is the insulation of the inner conductor. This insulation may be rubber or plastic. In some cables, the insulation around the core is wrapped with a packing material such as strands of cotton. These fibers provide some additional protection for the conductor.

Finally there are the wires of the core. In audio cables, the inner conductor or conductors are made up of fine stranded wire, for better flexibility and less likelihood of breakage.

Stripping the Cable

Before starting to strip the cable, place the boot and the strain relief onto the cable, as shown in figure 18-6. (Unlike the Switchcraft connector, the metal sleeve does not go onto the cable before soldering.)

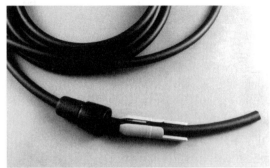

Figure 18-6.

To be able to attach the cable to the connector, you need to strip the cable appropriately. If too short a length at the end is stripped, it will be very difficult to solder, and the solder joint will be weak. If too much is stripped, the conductors may contact each other accidentally, causing a short or noisy cable. Stripping too far also interferes with the function of the strain relief.

The right tool for stripping the outer insulation is a craft knife. You want to avoid nicking or cutting through the fine copper wires of the shield. Hold the knife like a pencil. With your other hand, bend the end of the cable between your thumb and forefinger. Use the knife to slice through the insulation about 5/8" of an inch from the end of the cable, gently slicing down into the insulation only at the top of the bend where it's under tension and being pulled apart. You don't actually need to move the knife back and forth the way you do when slicing bread. Just apply the knife blade to the stretched surface. This is shown in figure 18-7. With care and practice, the edge of the knife will barely graze the surface of the shield wires.

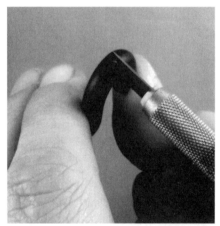

Figure 18-7.

Release the pressure on the bent cable end. Roll the cable about a quarter turn away from you, so that the slice in the insulation is on the side instead of the top. Bend the end of the cable again. Repeat the cut, starting in the existing cut and extending it toward you. This is shown in figure 18-8.

Continue this procedure until the outer insulation is cut all the way around. Remove the insulation.

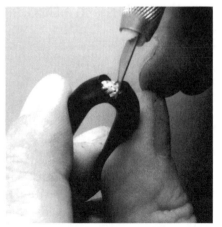

Figure 18-8.

Use a pointed reamer or other tool to unbraid the shield and pull it aside, revealing the inner insulation. Unbraiding is a process of gently pulling the reamer up through the strands of the shield to get them to disengage from each other. This is shown in figure 18-9. In a pinch, the tip of a multimeter probe can be used for unbraiding. If there is any cotton or other packing material, trim it off neatly.

The next step is part of the Zen of building cables. It's not essential, but following this practice will ultimately help you build cables faster, more accurately, and with less frustration. The essence of this step is to work with how the cable lies on the work surface. If you skip ahead and just continue on without this step, you will likely have to twist the cable into place to get it to stay, and this is harder and more frustrating than working with how the cable lies naturally.

Hold the connector body in the vise in the position where it is to be soldered, with the outer cup at the bottom. Coil the cable with the stripped end hanging out, setting it on the workbench in the same place that it will be when you are soldering. Hold the free end up to the connector in the vise, and observe which side is naturally on top when the cable is held up to the connector. This is shown in figure 18-10.

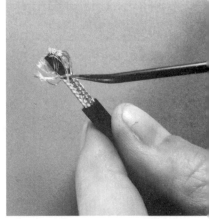

Figure 18-

Figure 18-1

Now that you know which side of the cable belongs on the top, twist the shield together on the opposite side, on the bottom. Use needle nose pliers to flatten the twisted shield, as shown in figure 18-11. The flattened shield will allow more room in the connector for the insulated inner wire.

Tinning the Cable and Connector

Turn on your soldering iron or station, and get its sponge wet. Test the hot iron with solder to see if it is tinned properly. A well-tinned soldering iron will hold a wet layer of melted solder on its tip.

The next step is to apply a light layer of solder to the shield. This tinning will hold the shield together so that it can be trimmed to length without the wires escaping. One of the goals of soldering a shielded cable is to keep all of the tiny wires that make up the shield under control, to prevent any of them coming in contact with the center cup or its wire. Having the shield already tinned also makes soldering the joint easier. This is shown in figure 18-12.

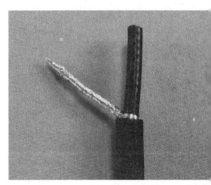

Figure 18-1

Figure 18-1

Tin the inside of the outer cup of the connector body while holding it in the vise, as shown in figure 18-13. An untinned body is shown for comparison. You don't need much solder here, just enough to wet the surface. The two key points of adding this solder are to be sure to heat the metal sufficiently so that the solder will melt properly when it is applied, and to use the usual sequence of iron in, solder in, solder out, iron out. It's not good soldering practice to apply solder to each part and simply melt them together to make the joint. Tinning both surfaces also reduces the chance of creating a cold joint. The layer of solder should be thin, smooth, and bright, and should extend over most of the inner surface of the outer cup. It's normal to see a little bit of brown or black residue at the edge of the solder. This is the remainder of the flux.

Figure 18-13.

Tin the inside of the center cup also. Be careful to just apply a thin layer of solder and spread it around inside the cup, so that the cup doesn't get filled and prevent the wire from fitting inside during later soldering. This is also shown in figure 18-13.

When the cable is completed, we should see the following, as shown in figure 18-14:

- The edge of the outer insulation is about 1/8" or less from the end of the outer cup.

- The edge of the inner wire's insulation is very close to the end of the center cup. This distance should be less than the thickness of a penny.

- The braided shield does not extend under the center cup, so that there is no chance that excess solder will come up and make contact between the shield and the center cup.

- The conductive sleeve has been stripped from part of the inner insulation to prevent it from accidentally contacting the center conductor.

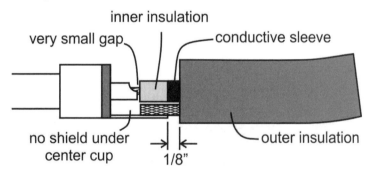

Figure 18-14.

Trim the end of the soldered shield short enough that it can't touch the underside of the center cup when it is in position in the outer cup. When it's soldered in place, it should fit underneath the cup in any case, but we'll be extra sure to avoid solder bridges between the shield and the center by trimming it shorter. An added advantage is that there will be less shield surface to solder. This is shown in figure 18-15. Positioning the cable as shown lets you know how far to strip the center wire.

Figure 18-15.

Use wire strippers to strip the insulation from the center wire. Be careful to use the right size set of notches in the stripper so that you just cut through the insulation without cutting into the wires inside. If you cut through one or two, there's no harm done. The wire should be stripped so that the inner insulation is even with the end of the shield, as shown in figure 18-16.

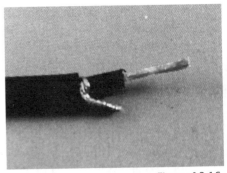

Figure 18-16.

Twist the center wires gently to keep them all together, and tin them. Use the knife to gently slice around the conductive sleeve just beyond the shield, as shown in figure 18-17. Try not to cut into the insulation underneath the sleeve. Remove the conductive sleeve. It's also possible to do this job with one of the larger sets of holes on the strippers, but the knife gives a more accurate and cleaner edge and is less likely to penetrate into the insulation underneath.

Figure 18-17.

Trim the center wires so that they just fit into the cup with the insulation touching the edge of the cup. This dry fitting is shown in figure 18-18, with the connector body held in the Panavise.

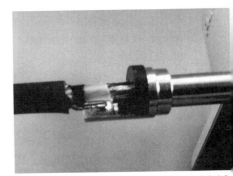

Figure 18-18.

Soldering the Connector

Hold the connector body firmly in the Panavise with the cups facing up. To solder the shield into the outer cup, bend the center wire back out of the way and use the Helping Hands to position the shield wire group in the cup. It may take some time to get it stable and positioned correctly, but it's worth the time, and it gets easier with practice. When you are starting out, it's normal for the positioning step to take longer than the soldering step. Remember that you can't hold anything in place by hand during soldering, because you need one hand for the soldering iron and the other hand for the solder.

Finally it's time to solder the shield into the outer cup. Be careful not to get any solder on the outside of the cup, because it will interfere with the fit of the sleeve. As with all soldering, you want the wire and the cup to get hot enough that the solder flows and spreads out over the inside of the cup. Get as much of the soldering iron tip in contact with the shield and the cup as possible. This is shown in figure 18-19.

Figure 18-19.

There is a tendency to not use enough solder, so make sure that you fill up the space around both edges of the shield while continuing to heat the joint. The pool of solder around the shield should be smooth and bright while it is melted, and it should remain smooth and bright afterwards.

Let the joint cool enough to handle it. Use needle nose pliers to thread the tip of the tinned center wire into the center cup and press it into place so that the wire doesn't form a hump. This is shown in figure 18-20. Use the Helping Hands jig to position the cable so that the center wire enters the cup in a straight line when viewed from the top and from the side.

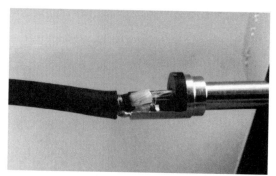

Figure 18-20.

Solder the inner wire into the cup. Be careful to get enough solder in the cup to make a good joint, but don't overheat the cup, because the plastic can melt. The solder should fill the cup and should be smooth and bright. This is shown in figure 18-21. Soldering this joint is much faster than soldering the outer cup, because there is much less mass to heat up. Let the joint cool so that you can touch it. Flex the cable to test the joints.

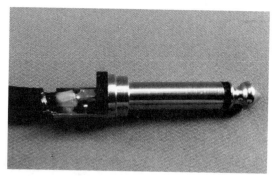

Figure 18-21.

Final Assembly

Slide the strain relief clamp up so that it mates with the outer cup. This is shown in figure 18-22.

Figure 18-22.

With the strain relief in place, slide the boot up around its end. This is shown in figure 18-23.

Figure 18-23.

Slide the sleeve over the body and screw it onto the boot. The soldered connector and cable should appear as shown in figure 18-24. There may be a small gap between the end of the sleeve and the edge of the wider part of the boot. If there is a large gap, as seen in the TS connector in the lower half of the figure, check that the spacer has been removed from the strain relief. If so, check that the sleeve is fitting

Figure 18-24.

smoothly over the body and the strain relief, and ensure that it is fully tightened. The sleeve should be able to be tightened and loosened by hand, with no tools.

Repeat the process for the other end of the cable. In particular, remember to put the boot and the strain relief on the cable before you start to strip the end. If you find that you have forgotten the strain relief, it can be installed later by spreading it open and sliding the cable into it. If you forget the boot on the second connector, you have no choice but to desolder the connector body and install the boot.

Testing the Cable

Having soldered together a cable, you need to check it to make sure it will work. You could use a cable tester, and every studio should have one available. You can also use a DMM, either functioning as a continuity tester or as an ohmmeter.

As you will remember from the discussion on testing fuses, continuity is the property of having an uninterrupted connection or path. A continuity tester is a meter or other device that tests whether two points are connected by a conductor. The meter probes are connected to the points, and a very low current is passed between them if there is continuity between them.

The DMM in figure 18-25 is set as a continuity tester. On this meter the continuity test setting is shown by a little musical note symbol and a diode symbol. Most manufacturers use a sound symbol to indicate that this setting has a tone, as shown in the meter in the previous chapter. The meter shown can also be used to check the polarity and function of a diode. Some other models of DMM have separate settings for testing continuity and diodes.

Figure 18-25.

When continuity is detected between the probes, most DMM models display **0;** otherwise the meter displays **1.** Some meters have other displays. Most models also include a tone that sounds when continuity is detected, so you don't have to be watching the meter display. When using the DMM as a continuity tester, always remember to check that the meter is working properly by touching the two probe tips together. While the tips are touching, the meter will change its display and give its tone if it has one.

There are three tests that you need to do to confirm that your TS cable is soldered correctly. When using the meter, make sure you are making good contact with the metal of the connector.

- Test between the tips of the two connectors. You should hear a beep, indicating continuity. This is shown in figure 18-26.

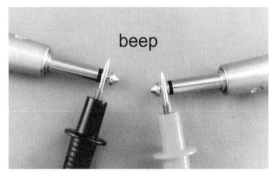

Figure 18-26.

- Test between the sleeves of the two connectors. You should hear a beep, indicating continuity. This is shown in figure 18-27.

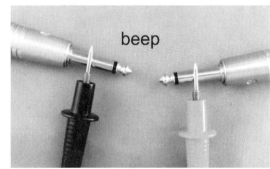

Figure 18-27.

- Test between the tip and the sleeve of either connector. You should hear no beep, indicating an open circuit with no connection. This is shown in figure 18-28. If the first two tests were positive, testing any combination of tip and sleeve will work, because you have already shown that the two tips are electrically equivalent, and the two sleeves are also electrically equivalent. A short between the center and the shield at one connector will test as a short at the other connector.

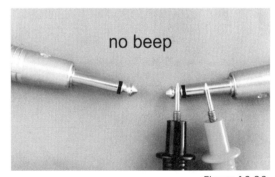

Figure 18-28.

If either of the first two tests fails, check the joints for failure. If everything looks all right, the only other conclusion is that the piece of cable is faulty. Before making this assumption, double-check that the meter is set correctly.

If the cable passes the first two tests but fails the third, then there is a connection making a short somewhere. The meter can't tell you which end to look at, because you've already shown that the tips are connected to each other, and the sleeves are also connected. Check your connectors for solder bridges, stray strands of wire, or anything else that causes a connection between the center wire and the shield wire.

For close-up examinations of solder joints, a stereo dissecting microscope is ideal, but they are expensive and heavy. A much more practical and portable method is a Coddington magnifier. This is similar to a jeweler's loupe mounted on a pen-sized flashlight, as shown in figure 18-29. It lights and magnifies the workpiece so you can see tiny pieces of stray wire, poor solder joints, and all sorts of other difficulties. The magnifier is also great for reading those tiny numbers printed on capacitors, transistors, and ICs.

Figure 18-29.

You should take a moment to congratulate yourself on finishing the TS cable. Neutrik connectors are challenging when you are starting out, so a completed cable that passes the tests and has smooth, bright solder joints is a cause for celebration.

19

The TRS Balanced Cable

Goal: When you have completed this chapter, you will have constructed a balanced audio cable with Neutrik TRS (tip-ring-sleeve) connectors.

Objectives

· Recognize the rationale for using balanced cable for professional audio connections.
· Strip and tin the shielded balanced cable properly for soldering.
· Tin the TRS connectors properly for soldering.
· Solder the connectors to the cable with clean, shiny solder joints.
· Assemble the cable's sleeve, boot, and strain relief appropriately.
· Test the cable for continuity, correct signal polarity, and shorts.

Building the TRS Cable

The TS cable is an example of an unbalanced cable, with only one conductor carrying the signal. The other conductor acts as a shield and as the common signal ground for the two devices being connected. While this is an effective cable in many situations, a balanced cable is the preferred method of connections between most pieces of professional studio gear. In this chapter, we will look at why balanced signal cables give better results, and we will build a balanced signal cable with Neutrik TRS connectors.

Materials

These are the materials you will need to build the balanced TRS cable with Neutrik connectors. Connector and cable specifications and sources can be found in the introduction.

– Two TRS connectors, model NP3C
– Balanced heavy-duty cable
– Safety glasses
– Soldering station or soldering iron
– Solder
– Helping Hands jig
– Panavise or similar vise
– Reamer/scraper tool
– Wire strippers
– Needle nose pliers
– Flush cutting wire cutters or side cutting wire cutters
– Craft knife (X-acto or similar)

Balanced Cables

A balanced cable consists of a total of three conductors, including two signal conductors surrounded by a shield. The way that a balanced cable is typically used is that the same signal is transmitted through the two conductors, except that one signal is 180° out of phase with the other. When these two signals arrive at their destination, the difference between them is amplified. This is equivalent to the out-of-phase signal being reversed again and the two signals then being added together. This cancels out much of the noise that may have been picked up along the way. This is shown in figure 19-1. Three types of connectors are typically used for balanced cables. The TRS (tip-ring-sleeve) connector is used for many line-level connections in the studio, and is a variant on the TS (tip-sleeve) connector. Both of these 1/4" plugs fit into the same size 1/4" jack, but only the TRS can carry balanced signal. XLR connectors are used on microphone cables and some line-level cabling. TT (tiny telephone) connectors are balanced connectors used for professional studio patch bays.

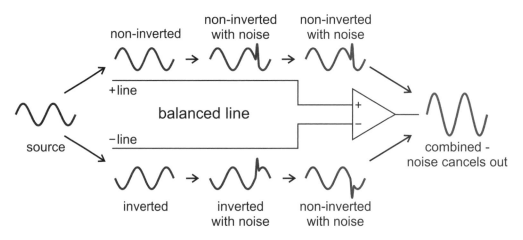

Figure 19-1.

The Neutrik TRS Connector

In some ways, the cable built with Neutrik TRS connectors is the hardest soldering project in this book. I suggest that you take your time with this connector, and don't be too concerned if you find that you need to take it apart and remake it. At the end you will be rewarded with a well-built cable that will give you reliable service for many years. The connector is shown as part of a cable in figure 19-2. The shaft of the connector has an additional band of metal between the tip and the sleeve. This band is called the ring. The overall length of the shaft is the same.

Figure 19-2.

The Connector Parts

The parts of the NP3C connector are essentially identical to the NP2C that you used for the unbalanced cable. These parts are the body, the sleeve, the strain relief, and the boot. The only difference is that the body has three solder contacts instead of two: the outer cup, the inner cup, and the ring tab. The parts are shown on the left in figure 19-3.

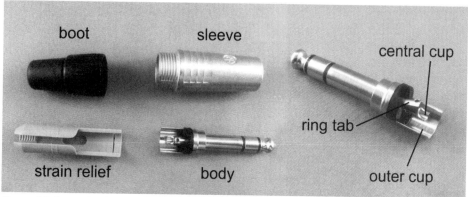

Figure 19-3.

The cable shield gets soldered into the inside of the outer cup. One signal wire gets soldered into the center cup, and the other one gets soldered to the ring tab. These parts are shown on the right in figure 19-3.

The strain relief is identical in design to the one on the TS connector. For the relatively thick cable you are using for this project, the extension ring is removed before putting the strain relief on the unstripped cable. For thin cable and medium cable, this extension is left in place. For cable that you are not familiar with, test the connector with the extension ring in place before deciding whether to remove it.

The Cable Parts

The cable you are using has two signal conductors with a shield. It is also called stereo cable, because we can also use it to carry two mono signals. One application is the cable for stereo headphones. The two strands of wire in the middle are for the signal, and the wrapping of wires around the outside is the shield. The layers of the cable are shown in figure 19-4.

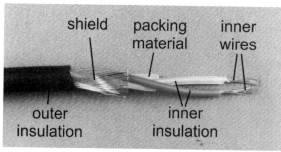

Figure 19-4.

On the outside is the outer insulation. This material is gripped by the strain relief and gives the cable its strength.

Inside the outer insulation is the shield. On the suggested Canare L-2T2S+ and Gepco M1042 cables, this shield is braided like the cable you used for the previous project. On the Mogami W2549 and Redco LO-Z1 cables, the shield is not braided. Braided shield is usually recommended, but Mogami is an industry icon, and many people have reported good results with their cable. I've been pleased with Redco's cable also, so braided shield doesn't seem to be essential for balanced cable.

The shield surrounds the conductors and some packing material. The packing is included so that the cable maintains a round shape even though it is made up of a pair of insulated conductors twisted together. Without the packing, the cable looks bumpy on the outside and is prone to developing flat spots. Each manufacturer chooses different packing materials and arrangements. Canare uses multiple strands of loose cotton fibers, Gepco uses cotton arranged in two cords, Mogami uses two strands of resilient plastic, and Redco uses two thick strands of jute.

The two conductors are what the packing materials are holding in place and protecting. Each conductor is protected by its own layer of insulation, which is color coded. There is no standardized color scheme. In the suggested cable types, Canare uses blue and white, Gepco uses black and red, Mogami uses blue and clear, and Redco uses blue and red. For the discussions that follow, blue or black are called "color 1" and are used to carry the (+) signal. White, red, and clear are called "color 2" and are used for the (−) signal. This is shown in table 19-1. Inside these color-coded insulators are the stranded wires that form the signal conductors.

cable type	color 1 (+) conductor	color 2 (−) conductor
Canare L-2T2S+	blue	white
Gepco M1042	black	red
Mogami W2549	blue	clear
Redco LO-Z1	blue	red

Table 19-1.

Stripping the Cable

Before starting to strip the cable, place the boot and the strain relief onto the cable. Stripping the outer insulation from the cable is just like the process you used for the TS cable. You need to strip the same length, about 5/8". This is about the width of your little finger. Hold the knife like a pencil, flex the cable, and slice through the outer insulation only, just barely grazing the shield. Remove the outer insulation. This is shown in figure 19-5.

Figure 19-5.

Use a pointed reamer to unbraid the shield and pull it aside, revealing the packing material and the two insulated inner conductors. On the Mogami and Redco cables, there is no unbraiding to do, so just unwind the shield and pull it gently aside. This is shown in the first panel of figure 19-6.

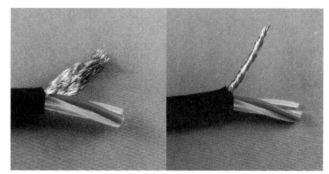

Figure 19-6.

With the TS cable, you learned to work with the position and "lay" of the cable. This technique is also helpful with the TRS cable. Another factor is that the color 1 wire needs to be soldered into the cup and the color 2 wire is soldered to the ring tab. This means that the ideal position for the shield to be gathered and twisted together is underneath the color 1 wire. This is shown in figure 19-7. Once the shield is twisted together, it needs to be squashed flat. This is shown in the second panel of figure 19-6. It's particularly important to flatten the shield on the TRS cable because the inner cup extends farther out, so the shield must slide underneath it with plenty of room.

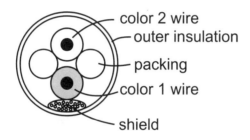

Figure 19-7.

Tinning the Cable and Connector

Prepare the soldering station for work by checking that the hot iron holds solder. As with the TS cable, the flattened shield needs to be tinned so that it will hold all of the wires together. Just a little solder will do the job here, as shown in figure 19-8.

Figure 19-8.

Trim the packing material neatly and as close as possible to the outer insulation. A pair of flush cutting wire cutters work well for this if they are sharp. For cotton or jute, a pair of small scissors also works well. This is shown in figure 19-9.

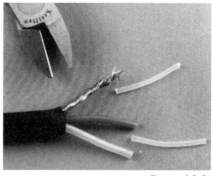

Figure 19-9.

Cut the shield to a length where it will slide 1/4"
into the outer cup when a 1/8" gap is left between the
outer insulation and the outer cup. This is shown in
figure 19-10. Notice that the color 1 wire is closer to
the shield than the color 2 wire.

Figure 19-

With the shield cut to length, the color 1 wire gets
stripped next. This wire will be soldered into the inner
cup. Strip it so that about 1/4" of insulation remains.
This is shown in figure 19-11.

Figure 19-

Gently twist the wires of the color 1 conductor together and tin them lightly. Cut the tinned wire group
to length so that the wire fits into the inner cup far
enough that there is only a very small gap between the
insulation and the cup. This is shown in figure 19-12.

Figure 19-

There is a hole in the ring tab, but we are not going
to use it. Instead the color 2 insulation gets stripped
so that it still extends over the inner cup, and the
color 2 wire gets soldered to the top of the tab. This is
shown in figure 19-13. With the insulation removed,
gently twist the wires and flatten them, as you did for
the shield. There is only a small amount of soldering
space on top of the ring tab, so the solder joint on top
needs to be kept as flat as possible.

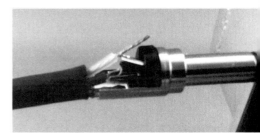

Figure 19-

Trim the color 2 wire to length and bend it down
over the ring tab, as shown in figure 19-14. This is
still a dry assemble step, without any soldering yet.
This is the last step for the cable before soldering, so
make sure that everything fits correctly.

Figure 19-

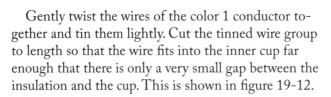

The outer cup, the inner cup, and the ring tab all need to be tinned before soldering the connector. The result is shown in figure 19-15. The solder from the tinning step should be thin, bright, and smooth. If it is lumpy, the solder has not been heated enough, or the iron tip may need to be checked to confirm that it is still wetting properly.

Figure 19-15.

Soldering the Connector

Hold the connector body firmly in the Panavise with the cups facing up. Bend the two signal wires back out of the way, and position the shield wire in the center of the outer cup with a 1/8" gap between the insulation and the edge of the cup. As before, this may take more time than soldering.

Make sure that the iron tip is in full contact with as much of the cup and the shield as you can reach without contacting the outer insulation. Once the tinned solder of the cup and the shield start to melt, adding solder to this melted mass is easier. Keep adding solder until you get a smooth, bright surface, but not so much that it connects with the cup above it. This is shown in figure 19-16.

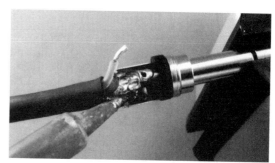

Figure 19-16.

When the body cools enough to handle safely, use the needle nose pliers to insert the color 1 wire into the inner cup. This is shown in figure 19-17. The cable will probably need to be supported so that it stays running straight into the connector. If it flops down, the color 1 wire will not stay in the cup.

Figure 19-17.

Solder the wire into place, filling the cup with bright, smooth solder. This is shown in figure 19-18.

Figure 19-18.

The remaining color 2 wire is a little trickier to deal with. It needs to be positioned on top of the ring tab, centered over the color 1 wire. One way to do this is to hold it in place temporarily with an alligator clip. This is shown in figure 19-19. A small piece of masking tape can also be used as a temporary clamp. Check that the wire is firmly in the correct position before starting to solder.

Figure 19-19.

Solder the color 2 wire to the ring tab. While you want enough solder to hold the wire firmly in place, avoid adding a big blob, as there is limited space between the ring tab and the inside of the strain relief when the connector is assembled. A side view of the completed cable is shown in figure 19-20.

Figure 19-20

The smooth, clear solder of all three joints is seen from another viewpoint in Figure 19-21.

Figure 19-21

Final Assembly

Slide the strain relief clamp up so that it mates with the outer cup, and then slide up the boot as shown in figure 19-22.

Figure 19-2

Slide the sleeve over the body and tighten it onto the boot. This tightening should not require any tools, just hand tightening. The completed connector and cable assembly are shown in figure 19-23.

Figure 19-2

With this end of the cable complete, repeat the procedure at the other end. Remember to slide on the boot and strain relief before you start stripping the cable.

Testing the Cable

As for the TS cable, the completed cable needs to be tested with a DMM set as a continuity tester. With two signal conductors and a shield, the number of tests has doubled. There are a total of six tests.

Three tests should give a positive result, showing continuity between the test points:

> Tip to tip
> Ring to ring
> Sleeve to sleeve

Three tests should give a negative result, showing no continuity:

> Tip to ring
> Tip to sleeve
> Ring to sleeve

To make sure that all the tests are performed, it is easiest to test each conductor at one end against all three at the other in turn. Only one of these tests should give a positive result. Here's the sequence: right probe on the tip, left probe on the tip (beep!), left on the ring (no beep), left on the sleeve (no beep), move the right probe to the ring, left on the tip (no beep), left on the ring (beep!), left on the sleeve (no beep), move the right probe to the sleeve, left on the tip (no beep), left on the ring (no beep), left on the sleeve (beep).

It's true that there is a little redundancy here. The advantage to doing it this way is that you really don't need to think about it too much. We'll see this redundancy again when we test the XLR cable.

When the tests are complete, this is cause for celebration. When you can solder Neutrik TRS connectors smoothly and accurately, you have shown yourself to be accomplished with a soldering iron. Most other soldering in the electronics world will seem fairly easy after making this cable. The XLR cable has a few new wrinkles for you, but to my mind the technicalities of the soldering are easier than those of the TRS cable.

20

The XLR Microphone Cable

Goal: When you have completed this chapter, you will have constructed and tested a balanced microphone cable with Neutrik XLR connectors.

Objectives
- Recognize the rationale for using star quad cable for balanced connections.
- Strip and tin the shielded star quad cable properly for soldering.
- Tin the XLR connectors properly for soldering.
- Solder the connectors to the cable with clean, shiny solder joints, maintaining the correct connections with the shield to pin 1 at each connector.
- Assemble the cable's sleeve, boot, and strain relief appropriately.
- Test the cable for continuity, correct signal polarity, and shorts.

Building the XLR Cable

The TS and TRS cables are both made with connectors that are based on the phone plug. In this chapter, we introduce the XLR connector and the star quad cable, which is a popular variant on the balanced cable used with the TRS connectors. The microphone cable with XLR connectors is the most common cable in professional audio.

Materials

These are the materials you will need to build the balanced TRS cable with Neutrik connectors. Connector and cable specifications and sources can be found in the introduction.

- One female XLR connector, model NC3FX
- One male XLR connector, model NC3MX
- Star quad cable
- Safety glasses
- Soldering station or soldering iron
- Solder
- Helping Hands jig
- Panavise or similar vise
- Reamer/scraper tool
- Wire strippers
- Needle nose pliers
- Craft knife (X-acto or similar)

XLR Connectors

The industry standard XLR connectors were developed by the Cannon company. There are several versions: 3-pin, 4-pin, 5-pin, 6-pin, and 7-pin. The 3-pin connector is the most common in audio applications. A 5-pin connector is used for DMX lighting control cables. Tube mics use 7-pin connectors, because they carry the high-voltage power supply for the tube and the low-voltage, high-current power supply for the tube's heater, as well as the balanced signal. The standard microphone connector configuration is shown in figure 20-1.

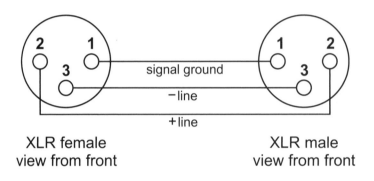

XLR female
view from front

XLR male
view from front

Figure 20-1.

Star Quad Cable

After developing the balanced cable, cable designers continued to experiment with cabling to achieve even lower noise levels. The result was four conductors arranged into two twisted pairs, called *star quad cable*. One conductor from each pair is used for the (+) signal, and the remaining conductors are used for the (–) signal. The double twisted-pair arrangement allows even more noise cancellation. Mogami claims that its star quad cable provides an improvement of 10–20dB over similar cables with a single twisted pair. There are numerous variations on the hyphenation and capitalization of the name. The terms *star quad*, *Star quad*, *Star Quad*, *Star-quad*, *Starquad*, and *Star-Quad* all refer to the same material. Most XLR and TRS cables are made with this 4-conductor cable. Cable with 2 conductors (as shown in the previous chapter) can also be used for these balanced connectors.

The outer insulation and shield are the same on the 2-conductor balanced cable and the single conductor cable. The two twisted pairs have the same color insulation on each, usually blue and white. The blue conductor from each pair is used for the (+) signal, and the white conductor of each pair is used for the (–) signal.

The Neutrik Female XLR Connector

The XLR cable is different from the cables you have built so far in that the connectors have different configurations. It is essential to connect pin 1 to pin 1, pin 2 to pin 2, and pin 3 to pin 3. If one isn't paying attention, it's common to switch pins 1 and 2, making a nonfunctional cable. We'll start with the female connector.

The Female Connector Parts

The female XLR connector is shown in figure 20-2. There is some potential confusion as to which XLR connector is denoted as female, because the female connector fits into an outer metal sleeve on the male XLR. That connector, in turn, has pins that fit into sockets in the female connector. The terms "male" and "female" refer to the pins and sockets, not to the outer sleeve.

Figure 20-2.

The parts of the female XLR NC3FX connector have similarities to the connectors you have used so far. There are four parts, as usual: the boot, the strain relief, the body, and the sleeve. These are shown in figure 20-3.

Unlike the TS and TRS strain relief systems, there is no adjustment in the XLR connectors for cable diameter. Thin cables and thick cables are all held equally well. When using thicker cable, there will be a gap between the boot and the sleeve so that it appears that the connector was not fully tightened.

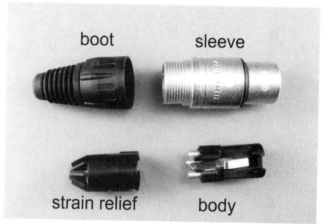

Figure 20-3.

The Cable Parts

Star quad cable has four wires in the middle, arranged in two pairs. These wires are wrapped in a shield and covered with the outer insulation. The parts of the cable are shown in figure 20-4.

The outer insulation and the shield are the same as for the cables you have used already. On the suggested Canare L-4E6S and Redco TGS-QD cables, the shield is braided from pretinned copper. The shield on the suggested Mogami W2534 is not braided and is bare copper, which Mogami claims provides a superior sound.

The shield surrounds the two pairs of conductors. The

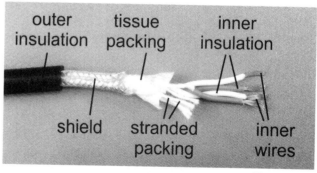

Figure 20-4.

Canare cable has a thin sheet of tissue paper wrapped inside the shield. In the Canare and Redco cables, there is also some packing material. The packing is included to fill the spaces around the twisted pairs of wires. Both Canare and Redco use multiple strands of loose cotton fibers.

The four conductors are in two pairs, with each conductor isolated by its own layer of color-coded insulation. With star quad cable, the industry has generally settled on blue and white for the insulation colors, but Mogami uses blue and clear. For these discussions, we will refer to these as blue and white. Inside these color-coded insulators are the stranded wires that form the signal conductors.

Stripping the Cable

Before stripping the cable, place the boot and the strain relief onto the cable, as shown in figure 20-5.

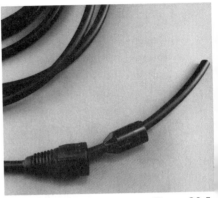

Figure 20-5.

Stripping the outer insulation is the same process you have used before. You need to strip the same length, about 5/8". This is about the width of your little finger. Hold the knife like a pencil, flex the cable, and slice through the outer insulation only, just barely grazing the shield. Remove the outer insulation. This is shown in figure 20-6.

Figure 20-6

As before, use the pointed tip of the reamer-scraper tool to unbraid the shield. This is a little more difficult than on the other cables, because the weave is tighter. Be patient and just hook into a few strands at a time, drawing the reamer toward the end of the cable. After a little while all of the wires will be unbraided. The example shows Canare cable, so there is a layer of tissue paper under the braid.

Figure 20-7

Before twisting together the wires, it's best to determine their position relative to how the cable is lying and which of the three cups in the connector body they need to be soldered into. The body is imprinted with raised numbers indicating the pin number. These imprints are on both the front or outside of the body, and on the back or inside. The front view is shown in figure 20-8. Pin 2 is on the left, pin 1 is on the right, and pin 3 is on the bottom.

Figure 20-8

These imprints are repeated on the back, as shown in figure 20-9. Here we are looking at the back of the connector body, so pin 1 is on the left, pin 2 is on the right, and pin 3 is on the bottom.

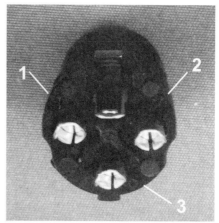

Figure 20-9.

For soldering, the body is held in the Panavise with pin 3 at the top, because this is the orientation where the cups are open for easy soldering. This puts pin 1 on the right, closer to us in figure 20-10. Pin 2 is on the left, farther away. This means that for optimal soldering, the shield wires need to be collected and pulled toward pin 1, the blue wires both need to be pushed away toward pin 2, and the white wires get pulled up to go to pin 3. This will make soldering easier when we get to that point, because the wires will already be in the right places.

Relative to where the cable is lying, all of the shield wires get pulled to the right, as shown in figure 20-11.

Figure 20-10.

Figure 20-11.

Twist the strands of the shield wire tightly together. Trim away the tissue wrap and the strands of packing material, as shown in figure 20-12. These should be trimmed down close to the outer insulation.

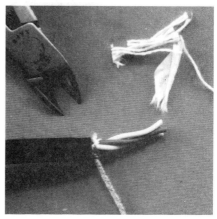

Figure 20-12.

Pull the blue conductors up and together, away from the shield. Pull the white conductors out away from the others. Getting the blues and whites in their proper places may require that they get untangled and rearranged where they exit the shield. This is shown in figure 20-13.

Figure 20-13.

Use the 20-gauge or 21-gauge hole on your strippers to strip a little less than half of the exposed insulation from the blue and white cables. The amount stripped should be identical on each wire, so that the same amount of copper is exposed on each. Be careful also to cut just the insulation, not the wires inside. This is shown in figure 20-14.

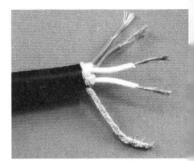

Figure 20-14

Twist the wires of each pair together, as shown in figure 20-15.

Figure 20-15

Tinning the Cable and Connector

Prepare the soldering station for work by heating it up and checking the tip. Use the soldering iron to tin these wires with a little solder, as shown in figure 20-16. As usual with tinning, not a lot of solder is needed, just enough to flow into the wires and hold them together. At the same time, tin about one-third of the end of the shield. You want the rest of the twisted shield to remain flexible, so don't tin more than this. Use minimal solder on the shield so that it will still fit into the cup.

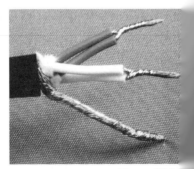

Figure 20-1

Trim the ends of the wires to length so that they fit into the three cups, as shown in figure 20-17. The insulation should come up very close to the cup. You want a good fit with the wire going to the bottom of the cup, so this step will probably involve taking several cuts to shorten the wire to just the right length. Take your time.

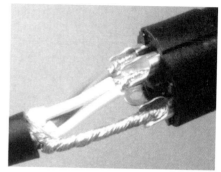

Figure 20-17.

Tin the insides of all three cups on the connector body. As with tinning the inner cups on the TS and TRS connectors, just a little solder is needed here so that the wires can still fit in easily. This is shown in figure 20-18.

Figure 20-18.

Soldering the Connector

Hold the connector body firmly in the Panavise with the cups facing up, as shown earlier in figure 20-17. To make the solder connections, use the Helping Hands jig to hold the cable in position with the blue wires in the cup for pin 2, the white wires in the cup for pin 3, and the shield in the cup for pin 1. Pin 2 is the first to be soldered, so check that it is fully placed in the cup with the insulation coming right up to the edge of the cup. Heat the cup and the tinned pair of wires with the tip of the soldering iron, and fill the cup with solder. Remove the iron and let the joint cool. This is shown in figure 20-19, with the cup at the bottom filled with bright, smooth solder.

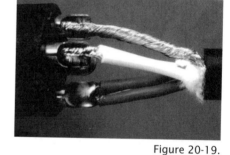

Figure 20-19.

Check whether the cable needs to be repositioned so that it extends in a straight line from the back of the connector body. Heat the shield and cup with the soldering iron and repeat the process of filling the cup. This joint will require more solder, because the wires of the shield will also soak up some solder. As before, keep adding solder until the cup is filled and the joint is bright and smooth. Remove the iron and let the joint cool. This is shown in figure 20-20.

Figure 20-20.

Repeat the soldering process with the cup for pin 3 and the white wire group. The result is shown in figure 20-21.

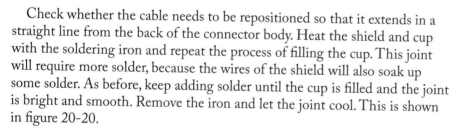

Figure 20-21.

Final Assembly

The strain relief needs to be properly aligned with the body for the sleeve to fit over both parts. Locate the half-round cutout in the rim of the strain relief, and align it with the metal tab with a hole in it, as shown in figure 20-22.

Figure 20-22.

Slide the strain relief up over the solder joints so that the cutout lines up and the edge of the strain relief touches the body, as shown in figure 20-23.

Figure 20-23.

The sleeve only slides onto the body in one place, so that the metal tab in the sleeve fits into the slot in the top of the body. This is shown in figure 20-24.

Figure 20-24.

Slide the sleeve over the body and the strain relief. When it is finally in place, only the cone-shaped fingers of the strain relief should be visible, as shown in figure 20-25.

Figure 20-25.

Slide the boot up and screw it onto the sleeve. With thick cable, you do not expect the gap between the outside of the sleeve and the outside of the body to close fully. If thinner cable is used, this gap will be smaller. This is seen in figure 20-26, which shows the completed cable.

Figure 20-26.

You have completed the female connector. The male connector at the other end of the cable is similar, but there are a few differences to pay attention to.

The Neutrik Male XLR Connector

The main difference between soldering the female XLR connector and the male XLR connector is that that pins 1 and 2 are reversed. It's useful to use the little rhyme "Pin 2 is blue" to remember to switch the positions of the shield and the blue wires.

The Male Connector Parts

The Neutrik XLR connector is shown in figure 20-27.

Figure 20-27.

There are four parts of the connector, as shown in figure 20-28. These familiar parts are the boot, the strain relief, the body, and the sleeve. The body consists of the three pins and their cups in a disc of black plastic. Be careful when soldering these pins, because overheating them can cause the plastic to soften, resulting in the pin changing its position.

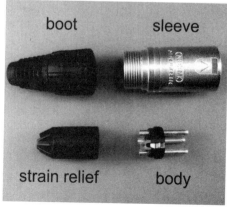

boot   sleeve

strain relief   body

Figure 20-28.

Stripping the Cable

Stripping the cable for the male connector is identical to stripping for the female connector, except that the blue wires and the shield are in opposite positions. In particular, you need to remember to place the boot and the strain relief on the cable before you start stripping. If you forget the boot, you will have to desolder and remove the connector to put it on the cable.

The male connector is marked on both sides with the pin numbers. The outside is shown in figure 20-29.

Figure 20-29.

The back of the body is shown in figure 20-30. Note that pin 1 is on the right as we view it from the back. When the body is turned upside down with the open tops of the cups upward for soldering, this places pin 1 on the left.

Figure 20-30

As before, strip 5/8" of the outer insulation from the end of the cable, and unbraid it with the reamer-scraper tool. Hold the connector body in the vise and coil the cable on the bench below it. Hold up the end of the cable to the connector to see which side the shield goes on. Gather the strands of the shield and pull them away from you, toward the cup for pin 1. Twist the strands together. Trim off the packing material, as shown in figure 20-31.

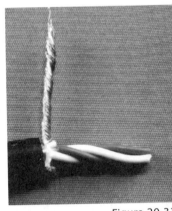

Figure 20-31

Spread out the conductors and pull them into their positions, with blue closer to you and white in the middle and on top. Just as for the female end, strip a little less than half the insulation from the ends of the wires, keeping the amount stripped even. This is shown in figure 20-32.

Figure 20-32

Tinning the Cable and Connector

Twist the stripped wires together and tin them. At the same time, tin about a third of the end of the shield, using a minimum of solder so that the shield will still fit into the cup. On the connector body, tin the insides of the three cups also, again using a minimum of solder. This is a fast procedure, but you should still be careful not to overheat the pins and soften the plastic block of the body.

Cut the tinned wires to length to fit into the cups with the insulation just coming up to the edge of the cup. Again, take your time to get a good fit with the wires going all the way into the bottom of the cup. This is shown in figure 20-33.

Figure 20-33.

Soldering the Connector

Hold the connector body firmly in the Panavise, with the cups facing up. This puts the cup for pin 3 at the top. To solder the wires in place, set up the cable with the tinned ends in their respective cups. Start with the cup for pin 2, with the blue wires. Work as quickly as is reasonable, but don't skimp on the solder. Fill the cup with bright, smooth solder. Let the connector cool down for a moment, then solder the shield into the cup for pin 1. After another brief cooldown, solder the white wires into the cup for pin 3. These three solder joints are shown in figure 20-34.

Figure 20-34.

Final Assembly

The strain relief must be aligned with the body for the sleeve to fit over both of them properly. The notch in the strain relief needs to be lined up with the little metal tab on the edge of the body, as shown in figure 20-35.

Figure 20-35.

The rectangular hole in the sleeve lines up with the notch in the strain relief, as shown in figure 20-36. When the body, the strain relief, and the sleeve are all lined up properly, the sleeve should slide on with little effort.

Figure 20-36.

With the sleeve on, only the tips of the strain relief should be visible, as shown in figure 20-37. Looking through the gap between the tips, the gripping surface of the strain relief can be seen to be resting on the outer insulation. This shows that the full cable is being held by the strain relief.

Figure 20-37.

Slide up the boot and screw it onto the sleeve. The completed connector is shown in figure 20-38. With this thicker cable, it's not expected that the boot will screw onto the sleeve far enough to close the gap.

Figure 20-38.

Testing the Cable

Testing the XLR cable is very similar to testing the TRS cable. There are three tests that should show continuity and three tests that should not. Testing the XLR is a little more difficult, because the contacts are harder to access. There are some tricks to make testing easier.

Three tests should give a positive result, showing continuity between the test points:

> Pin 1 to pin 1
> Pin 2 to pin 2
> Pin 3 to pin 3

Three tests should give a negative result, showing no continuity:

> Pin 1 to pin 2
> Pin 1 to pin 3
> Pin 2 to pin 3

The first test is shown in figure 20-39. The DMM is set as a continuity tester. The connectors are shown in the photo being held in a Panavise, but it is just as easy to hold them right next to each other in one hand. It is convenient to put one of the probes in pin 1 of the female connector to start with, and to let gravity hold it there while you move the other probe around to the pins of the male connector.

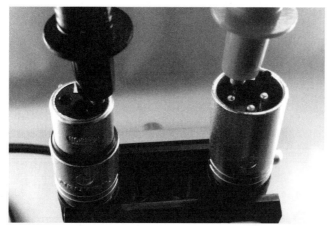

Figure 20-39.

Be careful to only make contact with the pin you are testing. Avoid unintentional contact with the other pins and with the metal sleeve that surrounds them. Remember that the connectors are mirror images of each other, so that the test shown in the figure is pin 1 to pin 1. While you need to do a total of six tests, I recommend doing a total of nine. Each time you move the probe to a new hole in the female connector, test each pin of the male connector. You should get only one positive result with each round of tests. If you make the full round of three tests each time, you know you've tested each one properly, and you don't have to think about which ones you have tested and which you haven't. This only takes an extra second or two, but the habit of doing so can help you avoid missing a short when you are tired or in a hurry. This will save you time in the long run.

If any of the continuity tests fail, check the joints for failure. If your solder is smooth and flows out into the cup as shown, this will never be a problem. If the cable shows any shorts, check your cables for solder bridges or stray strands of wire.

Congratulations on completing and testing the XLR cable. In some ways it is more difficult than the TRS cable because of the star quad cable, but in general I think it is easier. This is the last cable project in the book. The final project is a piezo transducer, which you can use as a signal source for listening to vibrating sound sources.

21

The Piezo Transducer

Goal: When you have completed this chapter, you will have constructed a piezo disc transducer connected to a light-duty unbalanced cable.

Objectives

- Recognize the rationale for using light-duty cable for the piezo disc.
- Strip and tin the shielded cable properly for soldering to the disc and connector.
- Tin the piezo disc properly for soldering without damage to the piezo material.
- Solder the disc and connector to the cable with clean, shiny solder joints configured for good strain relief.
- Assemble the cable's sleeve, boot, and strain relief appropriately.
- Test the function of the piezo transducer with an amplifier.

Building the Piezo Transducer

The last project is a *piezo disc*. This transducer can be used to pick up vibrations from anything that flexes as it creates sound, such as drum heads and guitar soundboards. The possibilities are nearly endless. The project consists of a cable with a 1/4" TS plug at one end and a piezo disc at the other. When completed, it can be plugged in and used as a signal source for recording.

Materials

These are the materials you will need to build the piezo disc project. Connector and cable specifications and sources can be found in the introduction.

- Piezo disc
- One TS connector, model NP2C
- Light-duty unbalanced cable
- Safety glasses
- Soldering station or soldering iron
- Solder
- Helping Hands jig
- Panavise or similar vise
- Reamer/scraper tool
- Wire strippers
- Needle nose pliers
- Craft knife (X-acto or similar)
- Masking tape

The Piezoelectric Effect

This project works on the piezoelectric effect. The word *piezo* is pronounced with three syllables, so that the first syllable rhymes with "bee," the second rhymes with "day," and the third rhymes with "go," with the stress on the second syllable: "pee-AY-zoe." The piezoelectric effect results from stretching or compressing certain crystal structures. When the crystals are bent, they generate a small electric field. This effect was first described in 1880 by two French brothers, Pierre and Jacques Curie.

Figure 21-1.

The brass disc shown in figure 21-1 is coated with a thin layer of piezoelectric crystal. When the disc flexes, the layer of crystal is alternately stretched out and squashed together. This action generates a voltage across the surface of the disc. A piezo disc can be used as a transducer to change vibrations into electrical signal, and it can also be used to change electrical signal into physical vibrations. The "beep" sound that a DMM makes to indicate continuity is produced by a piezo disc.

We will solder a conductor to the brass disc itself and another to the outer surface of the crystal so that we can amplify the signal being produced. The first step is to solder a TS connector to the other end of the cable, and the second is to solder the cable to the disc.

The layer of crystal on the disc is physically fragile and also sensitive to heat. A lightweight cable is recommended to reduce the stress on the solder joint with the piezo crystal surface, so that the piezo disc can withstand being used in more rigorous environments and will last longer. The shield will be soldered to the edge of the disc to reduce this stress also. We will also use some different soldering techniques to reduce the chance of damaging the surface with heat from the solder and iron.

The Neutrik TS Connector Revisited

The NP2C connector is already familiar to you from the TS cable. The only difference is that the cable for the piezo disc is thinner. The structure of this cable is familiar to you also, except that it is smaller. The cable layers are shown in figure 21-2.

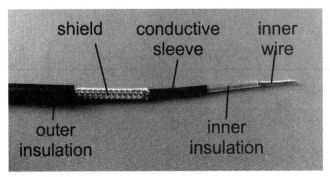

Figure 21-2.

Stripping and Tinning

As for the heavier cable, there is the outer insulation, the braided shield, the conductive sleeve, the inner insulation, and the inner wire. The procedure is very similar. One difference is that the spacer does not get removed from the strain relief of the TS connector, because the cable is so much thinner. Place the boot and the unmodified strain relief on the cable and slide them down. (If you forget and remove the spacer, all is not lost, but the strain relief will not be very effective, and any stress on the cable will be on the solder joints.)

Prepare your soldering station for work, checking that the tip is wetting properly. Strip 5/8" of the outer insulation, and unbraid the shield. The wires of the shield are thin and tightly woven, so it will take a little patience to unbraid them. Pull the shield wires together to one side of the cable. Twist them together tightly and tin them to hold them in a group. Trim the ends short enough that they don't slide under the inner cup of the connector. Hold the cable in place with the shield in the outer cup to see where to strip the inner insulation, as shown in figure 21-3.

Figure 21-3.

Strip the inner insulation to the appropriate length and tin the inner wires. Remove most of the conductive sleeve from around the inner insulation by slicing gently through it with a knife, as shown in Figure 21-4.

Trim the end of the tinned inner wire so that the insulation just touches the inner cup. Tin the inside of the outer cup. When this cools, tin the inside of the inner cup.

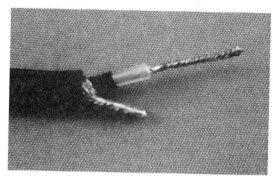

Figure 21-4.

Soldering and Assembly

Hold the connector body in the Panavise with the cups facing up. Bend the inner wire back out of the way and use the Helping Hands to hold the cable in position. Solder the shield into the outer cup. After the connector cools off, use needle nose pliers to place the inner wire in the inner cup. Solder the wire in place, filling the cup with bright, smooth solder, as shown in figure 21-5.

Figure 21-5.

Assemble the cable by sliding the strain relief up to meet the body and sliding the boot up over the fingers of the strain relief. Place the sleeve over the body and tighten it into the boot. The completed connector is shown in figure 21-6.

Figure 21-6.

The Piezo Disc

Soldering the piezo disc involves breaking some of the rules of soldering. In most situations it does not work to apply solder to the tip of the iron and then try to get the solder to stick to something else, because the flux has already been used up and because solder flows toward the iron. Because the disc is heat sensitive, heating the disc first and melting the solder onto it will destroy the crystal layer. Just this once, we will transfer solder from the iron to the work piece.

It's also not good practice to tin two pieces of metal and then melt their tinning together to get them to bond. Here we will do exactly that, because we don't want to overheat the crystals on the disc surface.

Some piezo discs are smaller than the one shown here. Some are coated on both sides, and some are just are just coated on one side. Either side of the double -coated discs will work for this project. If your disc is only coated on one side, use that side.

Stripping and Tinning the Cable

Strip off about an inch of the outer insulation, so that the stripped area is a little shorter than the diameter of the disc. This is shown in figure 21-7.

Figure 21-7.

Use the reamer-scraper tool to unbraid the shield. As at the other end, this may take some patience. Just unbraid a few strands at a time. The result is shown in figure 21-8.

Figure 21-8

Pull the shield all to one side and twist it together. Use the needle nose pliers to flatten the shield so that it lies flat on the surface of the disc, as shown in figure 21-9.

Figure 21-

Tin the shield on the half closer to the outer insulation. Use a knife to cut through the black conductive sleeve, being careful not to cut into the insulation underneath. This can be a little tricky because the sleeve material is very thin. Strip off about two-thirds of the inner insulation. Twist together the inner wires and tin them heavily. Usually tinning is a matter of adding just enough solder to wet the surface, but in this case the wires should be loaded up with as much solder as they can hold. This is because the joint between them and the piezo surface will be melted together later, so we want lots of solder to make the joint. This is shown in figure 21-10.

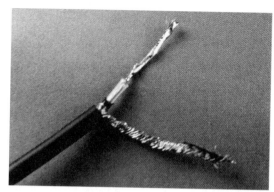

Figure 21-10.

Using the disc as a guide, trim the tinned shield and the tinned inner wires so that the stripped cable ends fit neatly in the circle of the disc, with the shield running along the exposed brass of the end and the inner wires in the area of the piezo crystal coating. This is shown in figure 21-11. There is nothing critical about where the inner wires get soldered to the piezo material, as long as they don't touch the brass of the disc at the edge.

Figure 21-11.

Tinning and Soldering the Piezo Disc

Note the positions of the two conductors shown in figure 21-12. You will do the same with your own disc and cable to determine where to tin the disc. Here I am envisioning a linear pad of solder in the position shown by the inner wire. Use masking tape to hold the edges of the disc down on the work surface. I've positioned the disc so that I can get a smooth sweep of the soldering iron across the surface to deposit the solder. For me, a diagonal sweep is the most comfortable, because my whole arm can be moved smoothly without trying to use my wrist. This is shown by the arrow in the figure. If another motion is more comfortable for you, that's fine. The goal is to move the iron smoothly across the surface. Try it with the soldering iron turned off to see what works best.

Figure 21-12.

Now it's time to break the soldering rules. With the iron hot, load up the tip of the iron with as much solder as it will hold without dropping off. It will probably take a few tries to get this right, so be prepared for some splashes of solder on your work surface. The loaded iron is shown in figure 21-13.

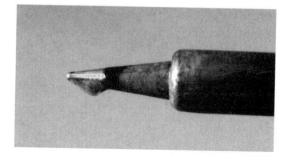

Figure 21-13.

Apply the solder to the disc in a smooth, gentle motion. If you move too slowly, the solder will stick but will create holes in the coating. Apply solder directly to the exposed brass at the edge with the soldering iron. The disc may deform slightly under the heat, but it will flatten out when it cools. These solder pads are shown in Figure 21-14.

Figure 21-14.

Tape the cable in place with the inner wire over the solder pad. Load up the soldering iron with solder again, and use it to melt the solder of the wire and the solder pad. When they have melted together, remove the iron.

Use the iron to heat the shield and its pad on the edge. When they melt, add more solder to complete the joint. The completed piezo disc is shown in figure 21-15.

Figure 21-15.

To test the piezo disc, plug it into an instrument-level input such as a guitar amplifier. The piezo disc is not a microphone, so it doesn't respond to sound directly. It only responds to being flexed or bent, so tapping on it will produce a signal. You are now ready to try it out on other vibrating sources.

Congratulations!

Congratulate yourself on completing all of the projects! These three cables and the piezo transducer form the basis for building a wide variety of cables, and they have given you the soldering skills to start working on the circuits of audio gear.

Afterword:
Pressing Onward

Thank you for taking the journey through this book with me. I hope that you have found it informative and inspiring.

Every piece of audio technology that we work on will present us with challenges. Some people seem to be able to zero in on the problem, and they never forget to reconnect a critical jumper wire before reassembling the unit. Most of us mere mortals need to work at it a little harder. It seems that the challenge is to remember that the process of building or fixing electronics isn't just about the end result; it's also about the journey of doing so.

It's also a challenge to remember that these pieces of gear didn't just fall out of the sky fully formed. They were designed by human beings, and they can be taken apart, diagnosed, and fixed by human beings. Most pieces of gear that need to be fixed have a lesson for us. It may be a lesson in humility, or persistence, or confidence, or recognizing when to stop, or perhaps learning to ask for help. Thanks to the Internet, it's relatively easy to find people who are struggling with similar problems or who know more than you do about them.

When you get to the end of a particular journey into the depths of a piece of audio gear, you can look back and consider what you have learned. Somewhere in the process, it probably helps to remember the words of the 18th-century explorer Juan Francisco De La Bodega Y Quadra: "I pressed on, taking fresh trouble for granted."

Prefixes

You are probably already familiar with numeric prefixes. We use them all the time when discussing the sizes of computer files: 220Mb (220 megabytes), 30Gb (30 gigabytes), and 2.3Tb (2.3 terabytes). We also talk about frequencies of 1000 Hertz and higher by using the kilo prefix, such as 2.5kHz (2.5 kilohertz).

In electronics we commonly use six prefixes to keep from needing to use a lot of zeroes. If we didn't use prefixes, we would need to distinguish between a 0.00000000033 farad capacitor and a 0.000000000033 farad capacitor. It's a lot easier to discuss the capacitors if we replace the zeroes with prefixes. This makes them 330 picofarad and 33 picofarad capacitors, which are easier to specify on a circuit diagram and easier to talk about.

The table in figure A-1 shows the two large prefixes and the four small prefixes that we commonly use and how to add and remove them.

prefix	symbol	meaning	to add prefix	to remove prefix
mega	M	× 1,000,000	move decimal six places left	move decimal six places right
kilo	k	× 1,000	move decimal three places left	move decimal three places right
milli	m	÷ 1,000	move decimal three places right	move decimal three places left
micro	m	÷ 1,000,000	move decimal six places right	move decimal six places left
nano	n	÷ 1,000,000,000	move decimal nine places right	move decimal nine places left
pico	p	÷ 1,000,000,000,000	move decimal twelve places right	move decimal twelve places left

Table A-1.

The ideal way to discuss voltages, capacitor values, currents, and other electronic quantities is to use numbers that are between 1 and 999. If the number is larger or smaller, we need to add a prefix. The second-to-last column shows how to add a prefix to a number that has too many zeroes. In the first example above, we need to move the decimal place to the right far enough to bring it into the 1–999 range. This is shown in figure A-1. In the same figure, we also see the process of adding a large prefix. Notice that the decimal is always moved in multiples of three.

$$0.000000000330 \text{ F} = 330\text{pF}$$

$$12,000 \ \Omega = 12\text{k}\Omega$$

Figure A-1.

The equations in the book don't work properly if the number has a prefix, so we also need to be familiar with removing prefixes and getting back the original number. Figure A-2 shows the process of removing a prefix from a small number, followed by an example of removing a prefix from a large number. Again, the decimal point is moved in multiples of three.

$$3.2\text{nF} = 0000000003.2\text{nF} = 0.0000000032 \text{ F}$$

$$2.5\text{M}\Omega = 2.500000 \text{ M}\Omega = 2,500,000 \text{ }\Omega$$

Figure A-2.

These processes of adding and removing prefixes are essential parts of working with the math in this book.

Appendix B: Solving Math Problems

I recommend using four distinct steps to solve any math problems. These are shown below. By using these steps consistently, you can solve any math problem. The idea is that when you list out the quantities that you know and the one that you don't know, the equation that you need will become apparent. The quantities you know can then be plugged into the equation and the answer calculated. If a prefix is needed for the result, you can then add one as described in Appendix A. This is the process that will be used to show the math in the book.

The Quantities, Equation, Solve, Prefix Process

Quantities: List the quantities that we know, such as voltage, current, and resistance. If there are any prefixes, we remove them, as shown in Appendix A.

Equation: Identify the equation we need, and rearrange this equation so we can solve for the quantity that we don't know.

Solve: Combine the quantities that we know with the equation, and solve the math. The resulting answer needs to have a unit.

Prefix: Add a prefix to the answer if one is needed, as shown in Appendix A.

Example of Q-E-S-P

In this example, we have been given the information that the voltage is 9V, the current is 25mA, and we want to know the resistance.

Quantities: We list the quantities that we know, and the one quantity that we're going to solve for. If there are any prefixes, we remove them.

$E = 9V$
$I = 25mA = 0.025A$
$R = ???$

Equation: Identify the equation we need, and rearrange this equation so we can solve for the quantity that we don't know. Here we know the voltage (E) and the current (I), and we said above that we want to calculate the resistance (R). Ohm's law is the equation that relates voltage, current, and resistance, so this is the one we will use. We need it rearranged to get R by itself on one side.

$$E = I \times R$$
$$R = E\ /\ I$$

Solve: Combine the quantities that you know with the equation, and solve the math.

$$R = 9V\ /\ 0.025A$$
$$R = 360\Omega$$

Prefix: Add a prefix if one is needed. Here we don't need a prefix, because 360 is already between 1 and 999.

$$R = 360\Omega$$

Modifying the DMM

The DMM's current measurement circuit is protected from excessive current flow by a small fuse. If more than about 250mA flows through the meter when measuring current, the fuse blows and stops the current. Unfortunately the fuse must then be replaced. Most people blow the fuse in their DMM when they are starting out, so it's more convenient to replace the fuse with a self-resetting circuit breaker, which is shown in figure C-1. Its part is item number TRF250-120-ND from DigiKey. This device detects excess current and disconnects the meter circuit automatically. When the danger has passed, it resets itself and reconnects the meter.

This procedure is well worth doing to reduce the frustration of working with the DMM while learning to measure current, but you should also be aware that making the modification will probably void the warranty of the meter. If you prefer not to modify the meter, I suggest that you obtain a supply of replacement fuses so that you can fix the meter relatively easily.

This modification only affects the circuitry for measuring current. Even with the fuse removed or blown, the meter's ability to measure resistance, voltage, and continuity are not changed.

Figure C-1.

Meter Modification Procedure

To modify your digital multimeter, follow the steps shown.

- If your meter has a protective plastic boot or cover, remove it as shown in figure C-2.

- Remove the back of the meter with the appropriate screwdriver. Inside, locate the fuse in its clips. In the example meter, it's along the side, as shown in figure C-3. In other meters the fuse may be somewhere else on the circuit board, but it will look similar. If your meter has two fuses, look for the smaller one.

Figure C-2.

- Remove the fuse. To do this, you may need to use a screwdriver or a pencil underneath the cylinder to pull it out of its clips.

- Examine the clips that the fuse fits into. Some have little prongs or tabs that fit into holes and are soldered on the underside of the circuit board. Others, such as the model shown here, are soldered directly to solder pads on the upper side of the board. If you can see the solder joint, it is easier to remove these clips to access the pads directly. If you can't see the solder joint, it is

Figure C-3.

easier to solder the fuse replacement into the metal of the clip. There is no functional difference between these two methods.

- To remove the clips with the visible solder joint, heat the clip with the soldering iron until the solder melts. Slide the clip sideways off the pad. If the clip sticks to the tip of the iron, pick it up with the iron and remove it by wiping the iron on the sponge of your soldering iron stand. If the clip doesn't stick, leave it on the circuit board until it cools and remove it with your fingers. Discard the clips. The cleared solder pads will now be visible, as shown in figure C-4.

Figure C

- Bend the leads of the self-resetting fuse out so that they are pointing away from each other, as shown in figure C-5.

Figure C

- Place the leads of the self-resetting fuse over the pads or inside the clips if they have not been removed. Heat the pad or clip and the lead with the soldering iron, and add enough solder to bridge any gap. Allow to cool, and repeat this for the second lead. Allow to cool. The self-resetting fuse is now installed, as shown in figure C-6.

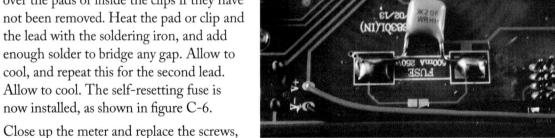

- Close up the meter and replace the screws, and put the meter back in its protective boot if it has one. You are now ready to

Figure C

use the meter. If you have the meter on the ammeter setting and you put the probes where you shouldn't, the meter will not register any current. The self-resetting breaker does not make any sound or other indication that it has disconnected the circuit. It may take a minute for the breaker to reset and reconnect the circuit if it has disconnected it.

Circuit Simulator

The circuit simulator is a piece of software that lets you build circuits and investigate their properties on a computer or tablet instead of out in the real world. The simulator has some distinct advantages. It's easy to get parts, because you just draw them. Nothing ever explodes or catches on fire or blows its fuse.

I've used a few simulators over the years, and the one I keep using is the Circuit Simulator v1.5h written by Paul Falstad. A screenshot of the program is shown in figure D-1.

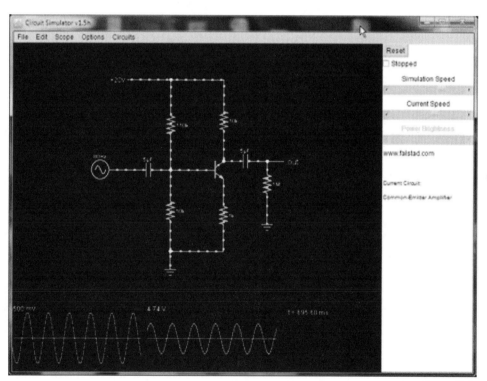

Figure D-1.

The simulator runs in Javascript, which runs on both PC and Mac. There is a big library of circuits to experiment with, and you can also build your own circuits. It's possible to save these new circuits as text files. The simulator lets you observe positive and negative voltages while showing you how much current is flowing at each point in the circuit.

Paul Falstad has graciously granted permission for me to include a copy of the program on the supplemental media. You can also obtain the software from his website at www. falstad.com, where it is listed in "math and physics applets" and then under the heading "analog circuit simulator applet."

There are other simulator programs and apps available. Among them is a system called SPICE, which many electrical engineers and circuit designers use for simulations. It is very accurate at predicting a circuit's behavior based on the precise qualities of the components. Perhaps because of this accuracy and precision, the SPICE programs I have used are not very user friendly, so I prefer Falstad's simulator.

Key Terms and Equations

A summary of the key terms, symbols, units, and definitions is shown in table E-1.

term	symbol	unit	unit symbol	definition
voltage	E	volt	V	the potential to do work
current	I	amp	A	the movement of electrons
resistance	R	ohm	Ω	the ability to restrict current flow
impedance	Z	ohm	Ω	resistance that changes with frequency
power	P	watt	W	the work being done by the current
capacitance	C	farad	F	the ability to hold an electric charge
inductance	L	henry	H	the ability to store energy in a magnetic field

Table E-1.

The formula circles and equations for Ohm's law and Watt's law are shown in figure E-1.

Ohm's law

$$E = I \times R$$
$$I = E \div R$$
$$R = E \div I$$

Watt's law

$$P = I \times E$$
$$I = P \div E$$
$$E = P \div I$$

Figure E-1.

The equations for total resistance and the voltage divider are shown in figure E-2.

Resistors in series $\qquad$ $R_t = R_1 + R_2$

general equation $\qquad$ $R_t = R_1 + R_2 + \ldots + R_n$

Resistors in parallel $\qquad$ $R_t = \dfrac{R_1 \times R_2}{R_1 + R_2}$

general equation $\qquad$ $\dfrac{1}{R_t} = \dfrac{1}{R_1} + \dfrac{1}{R_2} + \ldots + \dfrac{1}{R_n}$

Voltage divider $\qquad$ $E_{out} = E_{in} \times \left(\dfrac{R_2}{R_1 + R_2}\right)$

Figure E-2.

The equations for gain and the time constant t are shown in figure E-3.

Gain $\qquad$ $A_e = 20 \times \log\left(\dfrac{E_{out}}{E_{in}}\right)$

Time constant $\qquad$ $\tau = C \times R$

Figure E-3.

About the Online Files

The supporting online content contains five video segments that supplement the hands-on examples in the book. The first video segment provides clear specifics on the tools and techniques needed for getting started with soldering. The next four segments show how to build the four hands-on projects shown in the book.

Using the book, online materials (videos, supporting table of prefixes, electronics circuit simulator, and blank tables and graphs from the labs) will help you develop valuable electronics skills while reinforcing the concepts presented in the text.

Online Contents

Video Tutorials

1. Introduction to Soldering
2. Building the TS Cable
3. Building the TRS Cable
4. Building the XLR Cable
5. Building the Piezo Transducer

Additional Content

- Chapter 10. Blank Table and Graph
- Chapter 11. Blank Table and Graph
- Chapter 14. Blank Table and Graph
- Circuit Simulator Folder
- Circuit Simulator Supplemental Notes
- Quantities and Units
- Resistor Color Codes
- Table of Equations
- Table of Prefixes

Index